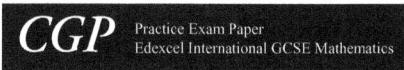

Edexcel International GCSE

Mathematics
Specification A
Higher Tier

Practice Set 1
Paper 1

Time allowed: 2 hours

Centre name				
Centre number				
Candidate number				

Surname
Other names
Candidate signature

In addition to this paper you should have:
- A pen, pencil and eraser.
- A calculator.
- A ruler.
- A protractor.
- A pair of compasses.

Tracing paper may be used.

For examiner's use			
Q	Mark	Q	Mark
1		13	
2		14	
3		15	
4		16	
5		17	
6		18	
7		19	
8		20	
9		21	
10		22	
11		23	
12		24	
Total			

Instructions to candidates
- Write your name and other details in the spaces provided above.
- Answer all questions in the spaces provided.
- In calculations show clearly how you worked out your answers.
- **You may use a calculator.**

Information for candidates
- There are 100 marks available for this paper.
- The marks available are given in brackets at the end of each question.
- You may get marks for method, even if your answer is incorrect.

Advice to candidates
- Work steadily through the paper.
- Don't spend too long on one question.
- If you have time at the end, go back and check your answers.

Exam Set MEHPI41

© CGP 2020 — copying more than 5% of this paper is not permitted

International GCSE Mathematics

Formula Sheet for Higher Tier

Arithmetic series

Sum to n terms, $S_n = \frac{n}{2}[2a + (n-1)d]$

Area of trapezium $= \frac{1}{2}(a+b)h$

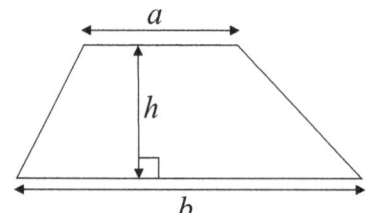

The quadratic equation

The solutions of $ax^2 + bx + c = 0$, where $a \neq 0$, are given by:

$$x = \frac{-b \pm \sqrt{b^2 - 4ac}}{2a}$$

Curved surface area of cone $= \pi r l$

Volume of cone $= \frac{1}{3}\pi r^2 h$

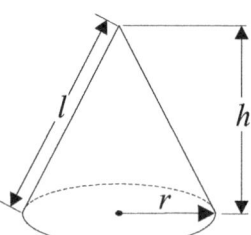

For any triangle ABC:

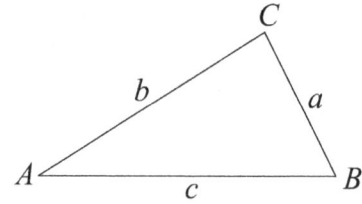

Sine rule: $\dfrac{a}{\sin A} = \dfrac{b}{\sin B} = \dfrac{c}{\sin C}$

Cosine rule: $a^2 = b^2 + c^2 - 2bc \cos A$

Area of triangle $= \frac{1}{2} ab \sin C$

Volume of sphere $= \frac{4}{3}\pi r^3$

Surface area of sphere $= 4\pi r^2$

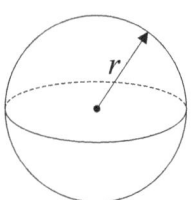

Volume of prism
= area of cross section × length

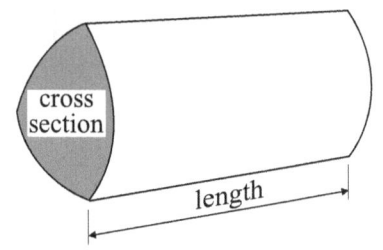

Volume of cylinder $= \pi r^2 h$

Curved surface area of cylinder $= 2\pi r h$

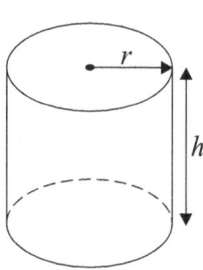

Answer ALL twenty-four questions.

Write your answers in the spaces provided.

You must show all of your working.

1   The first four terms of an arithmetic sequence are:

$$3 \quad 10 \quad 17 \quad 24$$

Write an expression for the nth term of this sequence.

....
[Total 2 marks]

2   (a) Fully simplify $\dfrac{12x^4 y^3}{2x^3 y^7}$

....
[2]

(b) Solve $3(x - 7) = 7x + 13$

$x = $ ....
[3]
[Total 5 marks]

3 Express $0.4\dot{1}$ as a fraction in its simplest form.

....

[Total 2 marks]

4 Declan keeps chickens and weighs all the eggs they lay.
The table shows the weights of eggs he collected last month.

Mass (*m*) in grams	Frequency
$40 \leq m < 50$	27
$50 \leq m < 60$	30
$60 \leq m < 70$	16
$70 \leq m < 80$	7

Calculate an estimate of the mean mass of Declan's eggs.

.... g

[Total 3 marks]

5 Show clearly that $16^{-\frac{3}{2}} = \frac{1}{64}$

[Total 2 marks]

6 The diagram shows the graph of $y = \sin x$ for $0° \leq x \leq 360°$.
The coordinates of the maximum point are $(90°, 1)$.

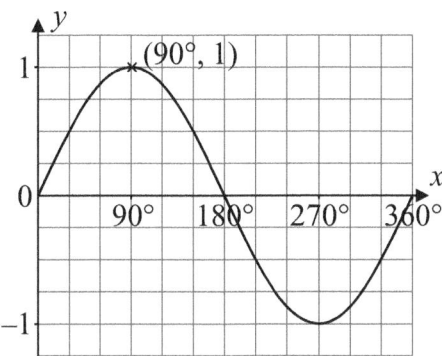

Write down the coordinates of the maximum point
of the graphs of the following functions for $0° \leq x \leq 180°$.

(a) $y = 3 \sin x$

....
[1]

(b) $y = \sin 2x$

....
[1]

(c) $y = 25 + \sin x$

....
[1]

[Total 3 marks]

7 The points *X* and *Y* are both above horizontal ground.
X is 10 m above the ground. The direct distance between *X* and *Y* is 70 m.

The angle of depression of *X* from *Y* is 32°.

Calculate the height of *Y* above the ground.
Give your answer correct to 1 decimal place.

.... m
[Total 4 marks]

8 Write $\frac{2}{7} - \frac{x+1}{x-3}$ as a single fraction in its simplest form.

....
[Total 3 marks]

9 The sets ξ, M and F are shown below.

ξ = {1, 2, 3, 4, 5, 6, 7, 8, 9, 10, 11, 12}
M = {multiples of 3}
F = {factors of 60}

(a) Complete the Venn diagram.

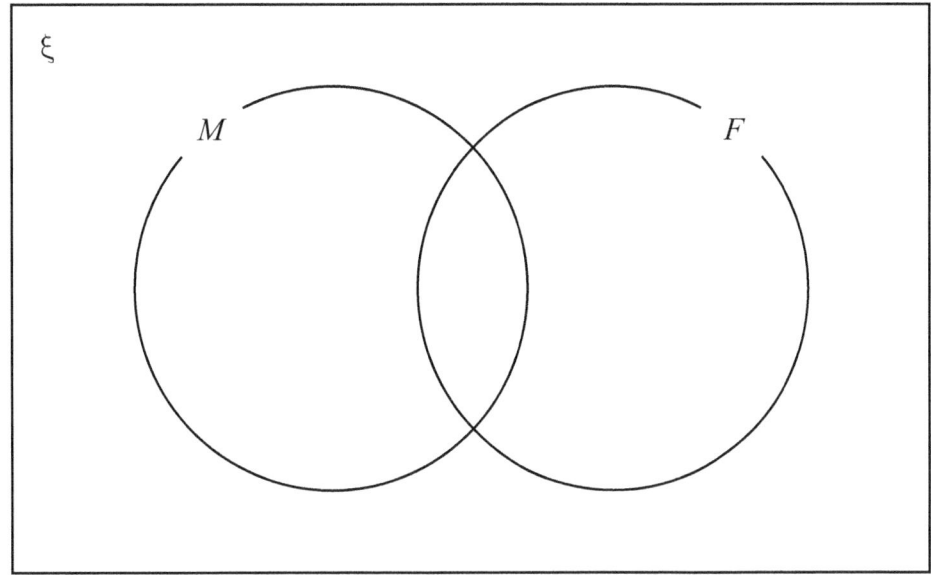

[3]

One of the numbers is chosen at random.

(b) Show that the probabilities of the number being in set (M ∩ F) or set (M ∪ F)' are equal.

[2]

[Total 5 marks]

10   Danilo gets a pay rise of 6%. His pay after the rise is €34 450.

(a) Calculate his pay before the rise.

€ ....
[3]

Aurea puts €7300 in an account that pays 2.5% compound interest each year.

(b) Calculate the amount in the account after 4 years.
Give your answer to the nearest euro (€).

€ ....
[3]

[Total 6 marks]

11   Show that the lowest common multiple of 450^3 and 240^3 is 60^6.

[Total 4 marks]

12 The graph of $y = f(x)$ is drawn below.
The line L_1 is a tangent to $y = f(x)$ at $x = 1$.

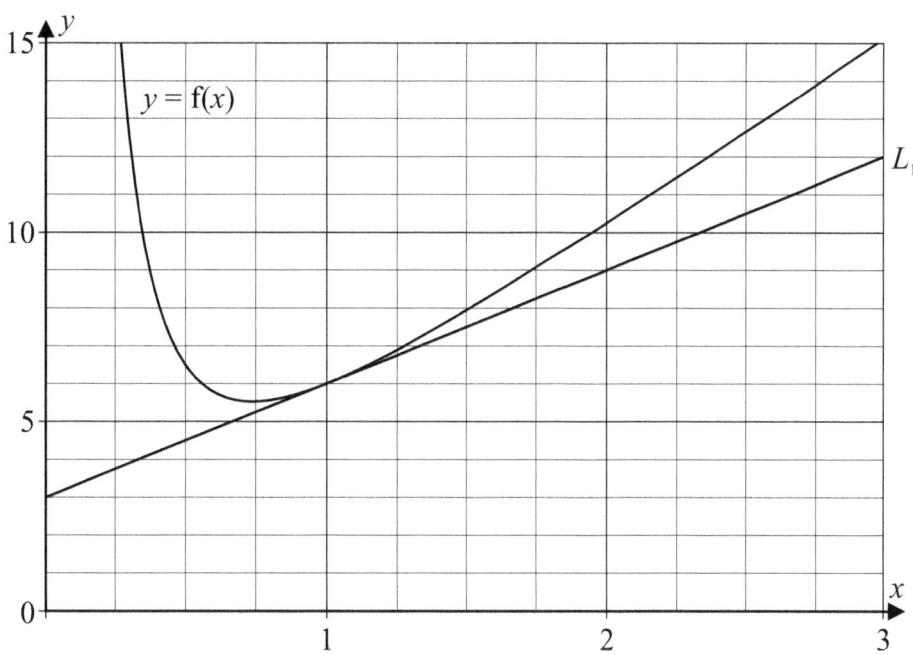

(a) Find the gradient of $y = f(x)$ at $x = 1$.

....
[1]

L_2 is parallel to L_1 and passes through the point $(-2, -7)$.

(b) Determine the equation of the line L_2.
Give your answer in the form $y = mx + c$.

....
[2]

[Total 3 marks]

13 The table shows the distribution of marks in a School Maths challenge.

Mark (m)	$m \leq 40$	$m \leq 60$	$m \leq 80$	$m \leq 100$	$m \leq 120$
Cumulative Frequency	6	20	50	68	80

(a) Draw a cumulative frequency graph to show these results.

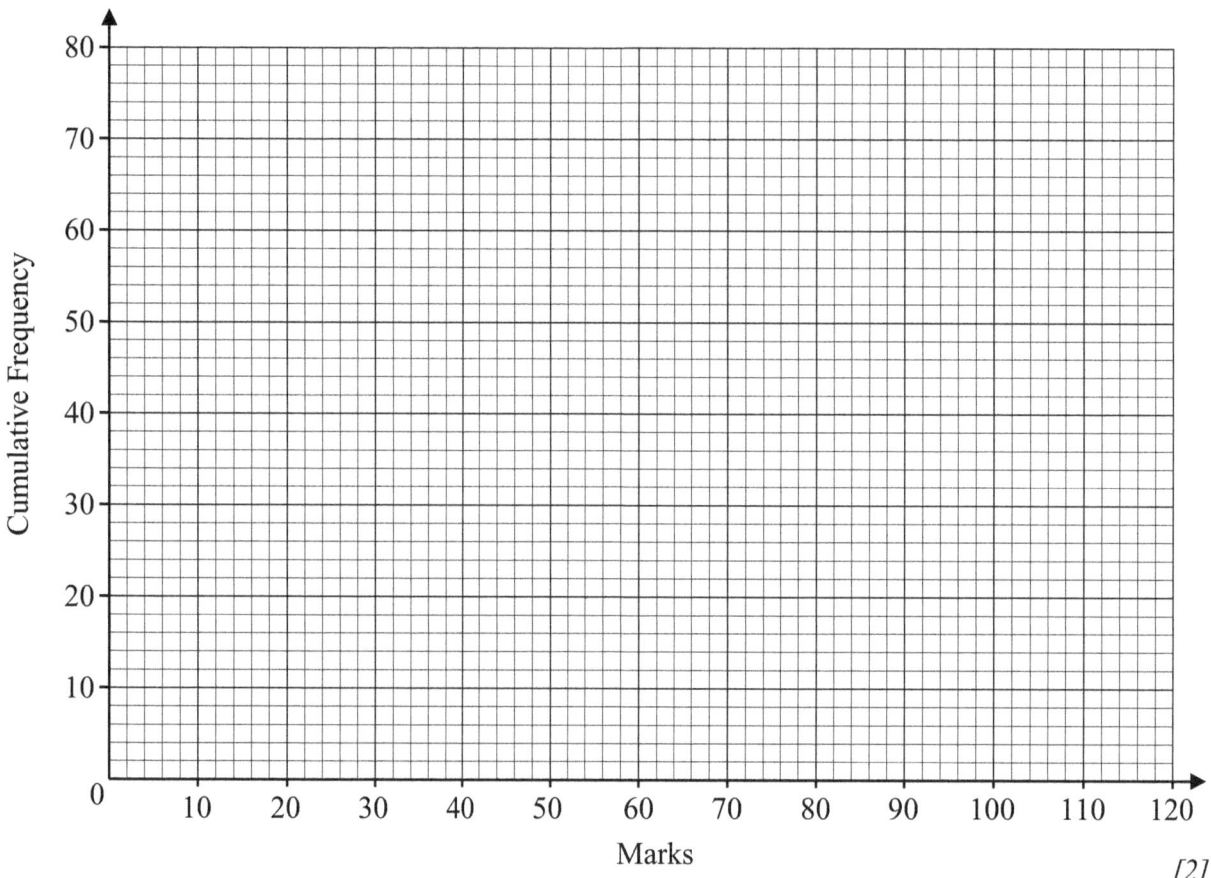

[2]

Students with 90 or more marks are awarded either a platinum or a gold certificate.
Platinum and gold certificates are awarded in the ratio 1 : 1.5.
Students with the highest marks are awarded a platinum certificate.

(b) Estimate the minimum mark needed to be awarded a platinum certificate.
Show how you get your answer.

....
[3]

[Total 5 marks]

14 Jane completes a 400 m wheelchair race in 64.5 seconds.
The distance is correct to the nearest metre and the time to the nearest half second.

Work out the lower bound of her average speed.

.... m/s
[Total 3 marks]

15 Solve the simultaneous equations

$$7x + 6y = 8$$
$$-x + 2y = 6$$

Show your algebraic working.

$x = $ ....

$y = $ ....
[Total 3 marks]

16 In a mosaic, the tile *TUVW* is a parallelogram.

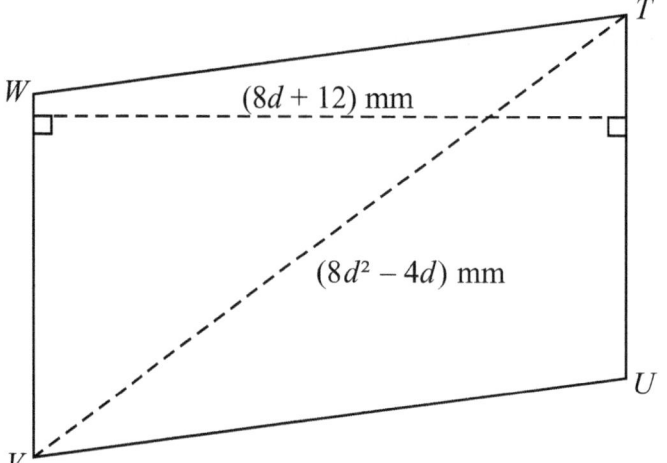

Diagram **not** accurately drawn

$TV = (8d^2 - 4d)$ mm and the width of the tile is $(8d + 12)$ mm.
The length of *TV* is *d* times the length of *VW*.

The area of the tile is 512 mm².

Find the length of *VW*.
Show how you get your answer.

.... mm

[Total 6 marks]

17 The two closed cylinders in the diagram below are mathematically similar.

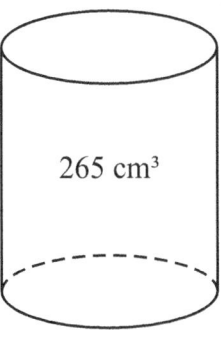
265 cm³

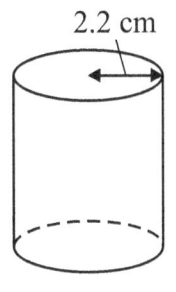
2.2 cm

Diagram **not** accurately drawn

The volume of the larger cylinder is 265 cm³.
The radius of the smaller cylinder is 2.2 cm.

The surface area of the larger cylinder is 3.6 times that of the smaller cylinder.

Work out the height of the smaller cylinder.
Give your answer correct to 1 decimal place.

.... cm

[Total 4 marks]

18 The diagram below shows two triangles, *ACB* and *ACD*.

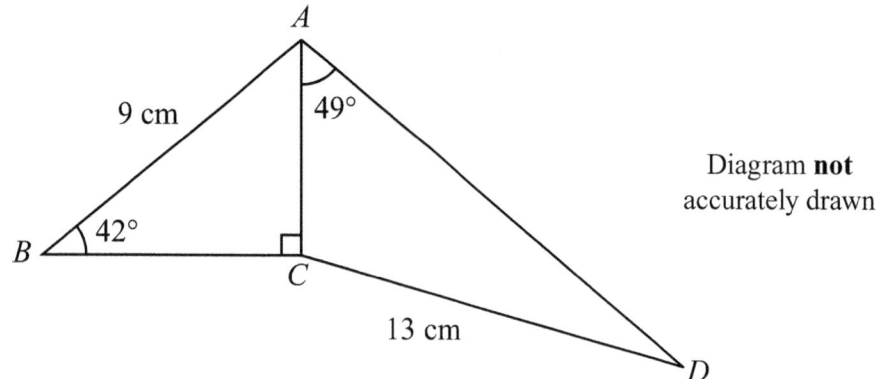

Diagram **not** accurately drawn

Find the size of angle *ADC*.

.... °

[Total 4 marks]

19 The functions f and g are defined as follows.

$$f(x) = 2x + 3$$
$$g(x) = f^{-1}(x)$$

(a) Solve the equation $(f(x))^2 = 4x^2$.

$x =$ ....
[3]

(b) Work out the value of $gg(x)$ when $f(x) = 27$.

....
[3]

The function h is defined as $h(x) = \dfrac{1}{\sqrt{f(x)}}$

(c) State the values of x that cannot be included in the domain of h.

....
[2]

[Total 8 marks]

20   The diagram shows a shaded circle drawn inside a sector of a larger circle, with centre O.

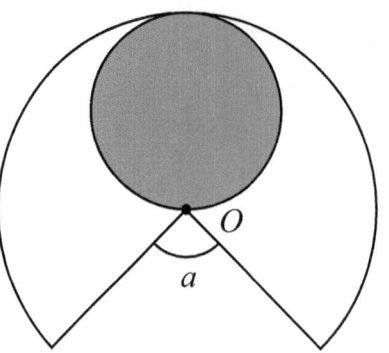

Diagram **not** accurately drawn

30% of the sector is covered by the shaded circle.

Find the size of angle a.

$a = $ .... °

[Total 4 marks]

21 The points *A*, *B* and *C* lie on a circle.
Point *O* lies at the centre of the circle.
Lines *DF* and *DE* are tangents to the circle at points *A* and *C* respectively.
Lines *AB* and *DE* are parallel.
Angle *BCE* = *n*

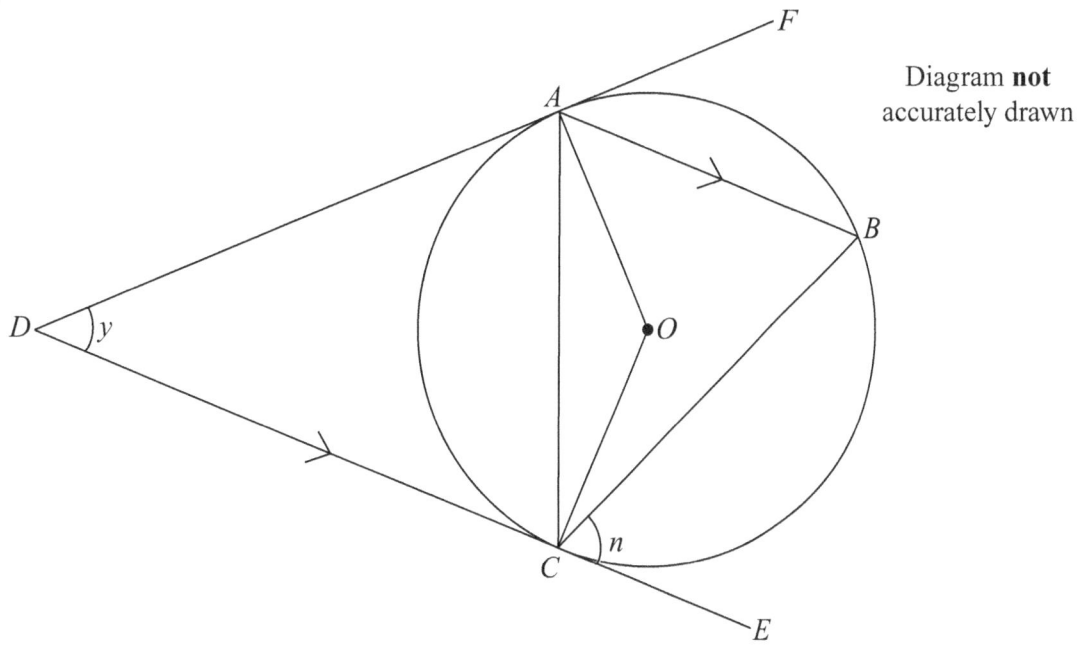

Diagram **not** accurately drawn

Show that $y = 180° - 2n$. You must give a reason for each stage of your working.

[Total 4 marks]

22 The diagram shows the parallelogram OABC.
The point D lies on AC, such that AD : DC = 3 : 2.
The point E lies $\frac{2}{3}$ of the way along line CB.

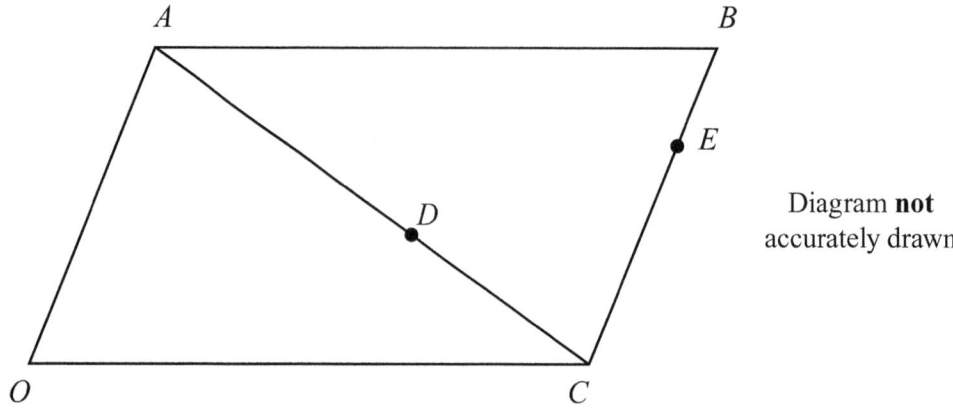

Diagram **not** accurately drawn

$\overrightarrow{OA}$ = **a** and $\overrightarrow{OC}$ = **c**

Show that ODE is a straight line.

[Total 5 marks]

23 A bag contains red beads and blue beads.
The probability of picking out a red bead is r.

One bead is picked out from the bag, its colour noted, and then it is replaced.
A second bead is then picked out.

The probability that exactly one of the beads is red is $\frac{4}{9}$.

(a) Find the possible values of r. Give your answers as fractions in their simplest form.

$r = $ .... and $r = $ ....
[4]

Isaac counts the number of beads in the bag and says,
"The number of blue beads and the number of red beads are both odd."

(b) Do you think Isaac is correct? Explain your answer.

....

....

....

....

[2]

[Total 6 marks]

24 Find the coordinates of both stationary points of the curve with the following equation.

$$y = x^3 + 3x^2 - 9x + 7$$

You must show your working.

.... and ....

[Total 6 marks]

[TOTAL FOR PAPER = 100 MARKS]

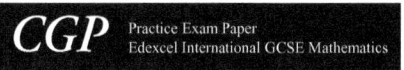

Edexcel International GCSE

Mathematics
Specification A
Higher Tier

Practice Set 1
Paper 2

Time allowed: 2 hours

Centre name				
Centre number				
Candidate number				

Surname

Other names

Candidate signature

In addition to this paper you should have:
- A pen, pencil and eraser.
- A calculator.
- A ruler.
- A protractor.
- A pair of compasses.

Tracing paper may be used.

Instructions to candidates
- Write your name and other details in the spaces provided above.
- Answer all questions in the spaces provided.
- In calculations show clearly how you worked out your answers.
- **You may use a calculator.**

Information for candidates
- There are 100 marks available for this paper.
- The marks available are given in brackets at the end of each question.
- You may get marks for method, even if your answer is incorrect.

Advice to candidates
- Work steadily through the paper.
- Don't spend too long on one question.
- If you have time at the end, go back and check your answers.

For examiner's use			
Q	Mark	Q	Mark
1		14	
2		15	
3		16	
4		17	
5		18	
6		19	
7		20	
8		21	
9		22	
10		23	
11		24	
12		25	
13		26	
Total			

Exam Set MEHPI41

© CGP 2020 — copying more than 5% of this paper is not permitted

International GCSE Mathematics

Formula Sheet for Higher Tier

Arithmetic series

Sum to n terms, $S_n = \frac{n}{2}[2a + (n-1)d]$

Area of trapezium $= \frac{1}{2}(a+b)h$

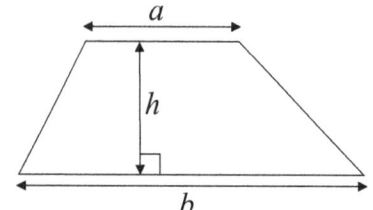

The quadratic equation

The solutions of $ax^2 + bx + c = 0$, where $a \neq 0$, are given by:

$$x = \frac{-b \pm \sqrt{b^2 - 4ac}}{2a}$$

Curved surface area of cone $= \pi r l$

Volume of cone $= \frac{1}{3}\pi r^2 h$

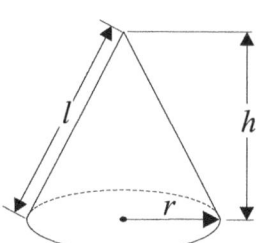

For any triangle ABC:

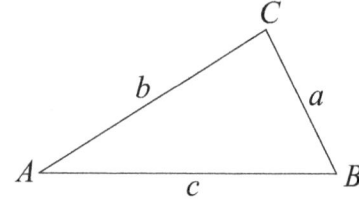

Sine rule: $\dfrac{a}{\sin A} = \dfrac{b}{\sin B} = \dfrac{c}{\sin C}$

Cosine rule: $a^2 = b^2 + c^2 - 2bc \cos A$

Area of triangle $= \frac{1}{2}ab \sin C$

Volume of sphere $= \frac{4}{3}\pi r^3$

Surface area of sphere $= 4\pi r^2$

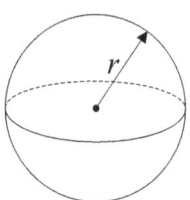

Volume of prism
= area of cross section × length

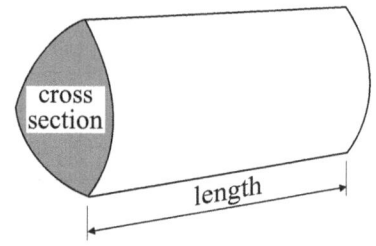

Volume of cylinder $= \pi r^2 h$

Curved surface area of cylinder $= 2\pi r h$

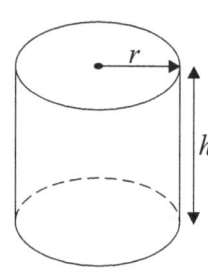

Answer ALL twenty-six questions.

Write your answers in the spaces provided.

You must show all of your working.

1   The number n expressed as a product of prime factors is $2^3 \times 3^2 \times 5$.

   Write n^2 as a product of prime factors.

   $n^2 = $ ....
   [Total 1 mark]

2   **a** and **b** are column vectors such that $\mathbf{a} = \begin{pmatrix} 8 \\ 3 \end{pmatrix}$ and $\mathbf{b} = \begin{pmatrix} 1 \\ -7 \end{pmatrix}$. Calculate:

   (a) $3\mathbf{a}$

   ....
   [1]

   (b) $\mathbf{a} - 4\mathbf{b}$

   ....
   [2]

   (c) the exact magnitude of $\mathbf{a} + \mathbf{b}$

   ....
   [2]
   [Total 5 marks]

3   The sets ξ, P and Q are shown below.

ξ = {positive integers less than or equal to 20}
P = {prime numbers}
Q = {1, 2, 3, 4, 6, 8, 12}

(a) List the members of the set $P \cap Q$.

....
[2]

(b) Find n($P \cup Q$).

....
[2]
[Total 4 marks]

4   Alison, Beckie and Che shared a lottery win in the ratio 7 : 3 : 2.

If Beckie's share was £11 367, how much more did Alison get than Che?

£ ....
[Total 3 marks]

5 *A*, *B* and *C* are points on a coordinate grid. *CAB* is a straight line.

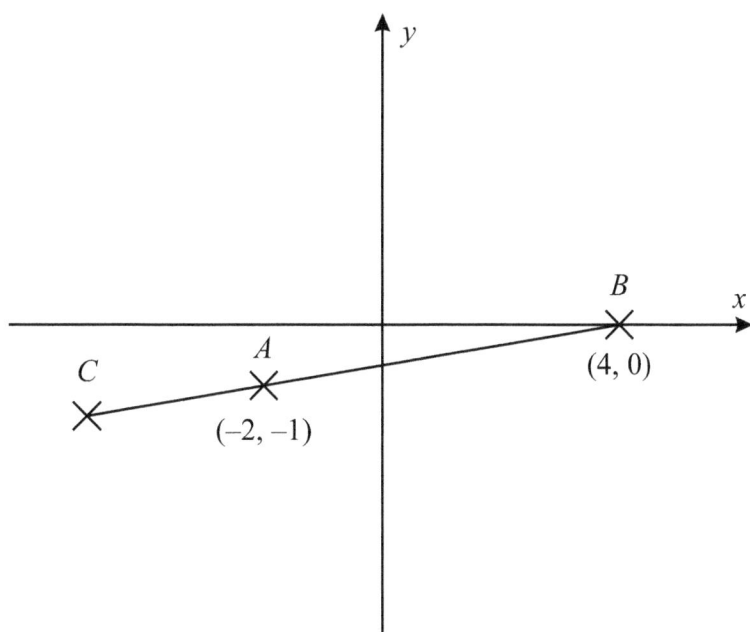

(a) Write down the column vector that translates point *A* onto point *B*.

....
[1]

AB is twice the length of *AC*.

(b) Use your answer to part (a) to write down the column vector that translates point *A* onto point *C*.

....
[1]
[Total 2 marks]

6 The list below shows the 15 values in a data set.
The data value *m* is the median.

    37 41 35 36 37 45 27 38 *m* 43 34 26 45 35 42

Find the interquartile range of the data.

....
[Total 3 marks]

7 The first three patterns of a sequence are shown below.

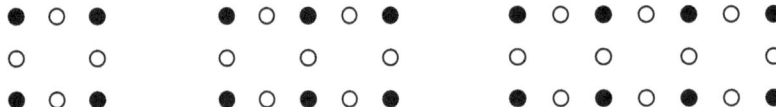

The sequence is continued until there are 1000 patterns.

How many white dots will there be in the sequence altogether?

....
[Total 3 marks]

8 Find the perimeter of the sector shown below.
Give your answer correct to 3 significant figures.

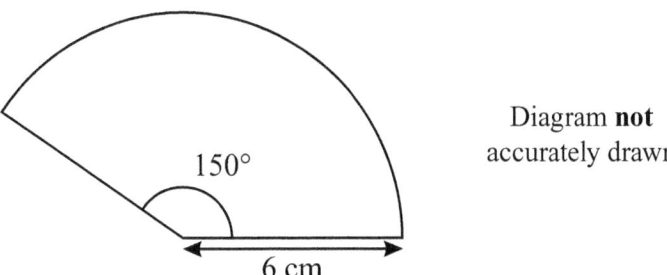

Diagram **not** accurately drawn

.... cm

[Total 3 marks]

9 Anton has some British pounds and Mexican pesos.
The exchange rate is £5 = 138 pesos.

Anton exchanges all of his pounds for pesos.
Afterwards, he has 5543 pesos altogether.

He has increased the number of pesos he originally had by 20.5%.

How many pounds worth of pesos did he have before the exchange?
Give your answer to the nearest whole pound.

£ ....

[Total 4 marks]

10 Make x the subject of $2y = \dfrac{3x}{2 - 5x}$

[Total 3 marks]

11 Here is triangle XYZ. Angle XYZ is acute.

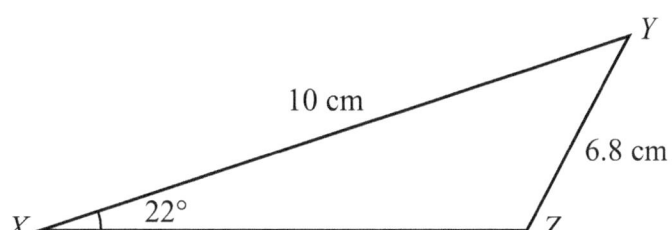

Diagram **not** accurately drawn

Work out the size of angle XZY.
Give your answer correct to 2 decimal places.

.... °

[Total 4 marks]

12 Expand and simplify $(x+3)(x+5)(x-2)$

....

[Total 3 marks]

13 $F = rs - 2t$

$r = 50$ correct to 2 significant figures
$s = 4.1$ correct to 2 significant figures
$t = 0$ correct to the nearest integer

Calculate the lower bound of F. Show your working.

....

[Total 3 marks]

14 The time taken for students to travel to school was recorded. The histogram shows the results.

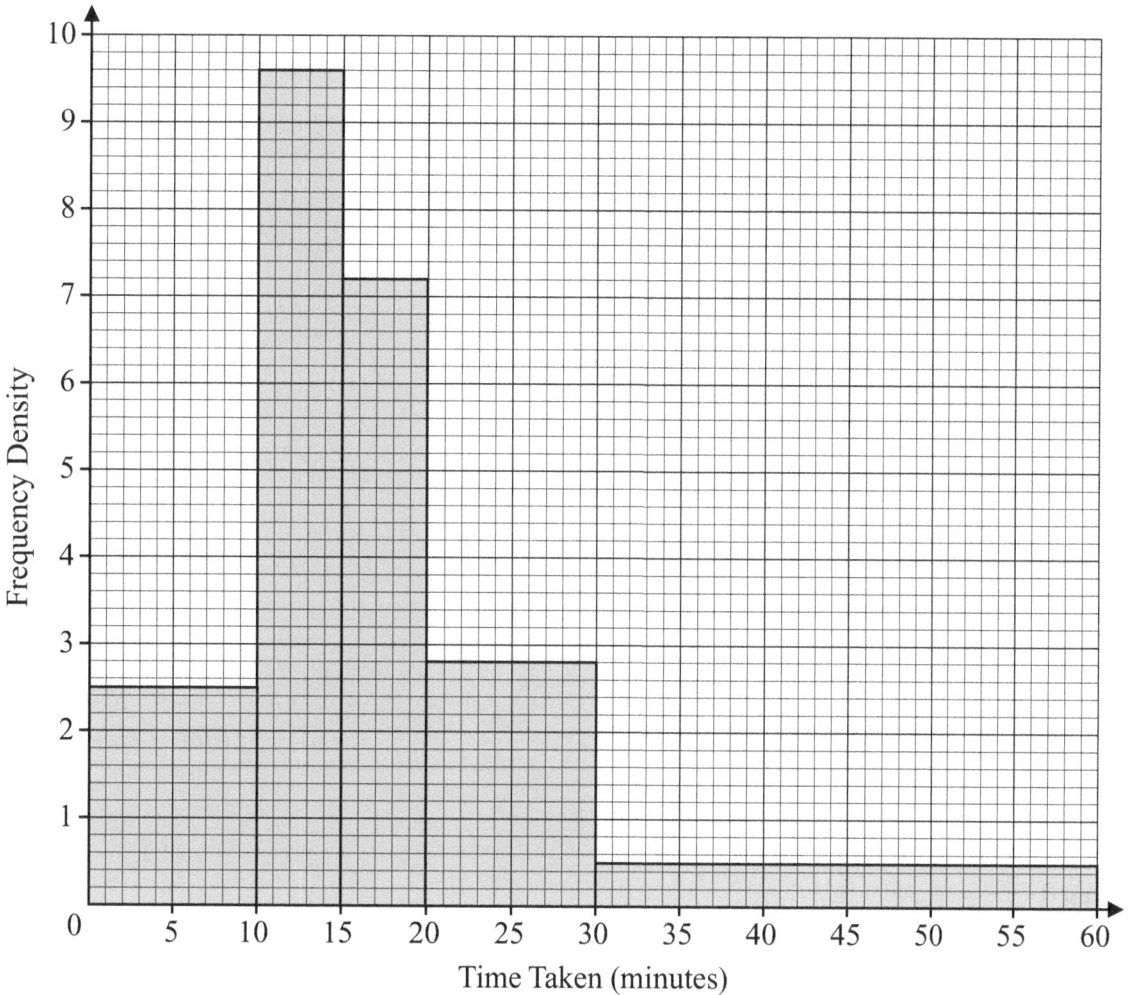

(a) Which class has the greatest frequency? Show your working.

....
[2]

The school day begins at 8:40 am.

(b) Estimate the number of students who must leave home before 8:15 am if they are to arrive at school on time.

....
[3]

[Total 5 marks]

15 A metal cone has a height of 4.3 cm and the radius of its base is 1.2 cm.
The mass of the cone is 17.5 g.

Work out the density of the cone in g/cm³.
Give your answer correct to 3 significant figures.

.... g/cm³

[Total 3 marks]

16 Show that $\dfrac{4}{3+\sqrt{5}} + \sqrt{5} = 3$

[Total 3 marks]

17 The mean of the numbers $3x$, 7, 13 and $\frac{y}{3}$ is 8.
The range of the numbers is 16.
$3x$ is a negative integer.

Find the values of x and y.

$x =$ ....

$y =$ ....

[Total 5 marks]

18 y is inversely proportional to the square root of x.
When $y = 12$, $x = 0.09$.

Find the value of y when $x = 0.16$.

$y =$ ....

[Total 3 marks]

19 The table shows the population and area of four countries.

Country	Population	Area (km²)
Afghanistan	3.80×10^7	6.52×10^5
Austria	8.96×10^6	8.39×10^4
Malaysia	3.19×10^7	3.30×10^5
Saudi Arabia	3.43×10^7	2.15×10^6

(a) Which two countries are closest in terms of their population?
You must show your working.

....

....

....

....
[2]

Population density is measured as the number of people per square kilometre.

(b) Which country has the greatest population density?

....
[2]

[Total 4 marks]

20 The spinner below is fair.
Priti spins it twice to produce a two-digit number.

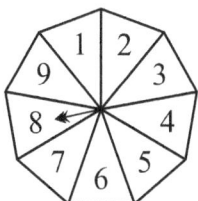

The digit shown on the first spin is the first digit in her number.
This digit is replaced on the spinner with 0.

The digit shown on the second spin is the second digit in her number.

What is the probability that her two-digit number is a multiple of 5?

....
[Total 3 marks]

21 A regular polygon has *n* sides and interior angles of 160°.

Find the size of each interior angle in a regular polygon with 4*n* sides.

....°
[Total 3 marks]

22 A curve has the equation $y = -5x^3 + \dfrac{x}{2} - \dfrac{13}{x}$

(a) Find $\dfrac{dy}{dx}$

$\dfrac{dy}{dx} = $ ....
[2]

The point *A* lies on the curve.
The *x*-coordinate of *A* is 1.

(b) Find the equation of the line that is perpendicular to the curve at the point *A*.
Write your answer in the form $ax + by + c = 0$, where *a*, *b* and *c* are integers.

....
[5]
[Total 7 marks]

23 (a) Solve the simultaneous equations

$$x^2 + y^2 = 20$$
$$x - 3y = 10$$

x = ...   y = ...

x = ...   y = ...
[5]

(b) How many points of intersection are there for the graphs with equations $x^2 + y^2 = 20$ and $x - 3y = 10$? Explain your answer.

....

....

....
[1]

[Total 6 marks]

24 Solve the following inequality.

(a) $x^2 - 1 \leq 3(x+3)$

....
[4]

(b) Show your answer to part (a) on a number line in the space below.

[1]
[Total 5 marks]

25 *OABCD* is a square-based pyramid.
The vertex *O* is vertically above the centre of the horizontal base *ABCD*.
OA = 12 cm and *AD* = 8 cm

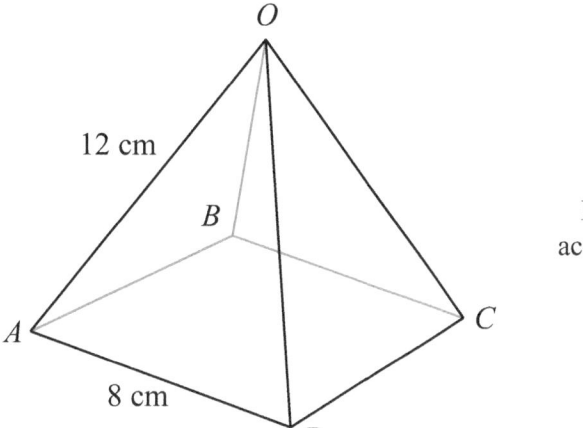

Diagram **not** accurately drawn

(a) Find the vertical height of the pyramid.
Give your answer in the form $a\sqrt{b}$, where *a* and *b* are integers.

.... cm
[3]

(b) Find the angle that the line *OA* makes with the plane *ABCD*.
Give your answer correct to 3 significant figures.

.... °
[3]

[Total 6 marks]

26 E, F, G and H are points on a circle.
HGK and $EJFK$ are straight lines. $EJ = FK$.

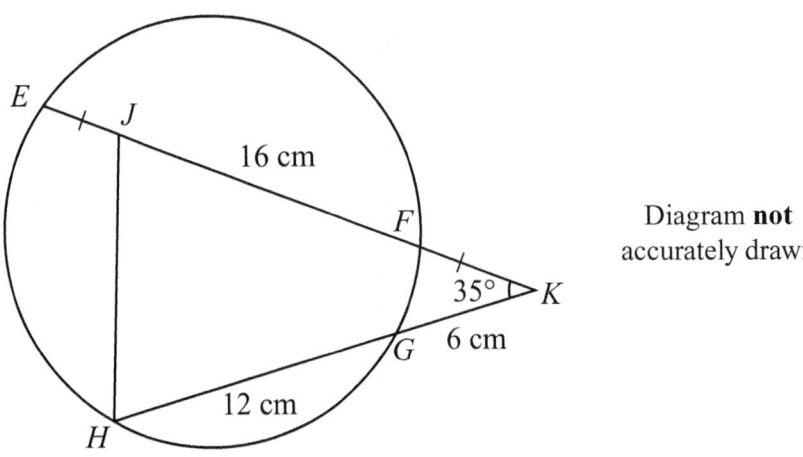

Diagram **not** accurately drawn

$GK = 6$ cm, $HG = 12$ cm, $FJ = 16$ cm and angle $JKH = 35°$.

Find the area of triangle HJK.
Give your answer correct to 3 significant figures.

.... cm²

[Total 6 marks]

[TOTAL FOR PAPER = 100 MARKS]

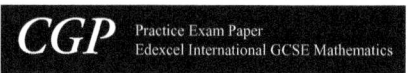

Edexcel International GCSE

Mathematics
Specification A
Higher Tier

Practice Set 2
Paper 1

Time allowed: 2 hours

Centre name	
Centre number	
Candidate number	

Surname	
Other names	
Candidate signature	

In addition to this paper you should have:
- A pen, pencil and eraser.
- A calculator.
- A ruler.
- A protractor.
- A pair of compasses.

Tracing paper may be used.

For examiner's use			
Q	Mark	Q	Mark
1		13	
2		14	
3		15	
4		16	
5		17	
6		18	
7		19	
8		20	
9		21	
10		22	
11		23	
12			
Total			

Instructions to candidates
- Write your name and other details in the spaces provided above.
- Answer all questions in the spaces provided.
- In calculations show clearly how you worked out your answers.
- **You may use a calculator.**

Information for candidates
- There are 100 marks available for this paper.
- The marks available are given in brackets at the end of each question.
- You may get marks for method, even if your answer is incorrect.

Advice to candidates
- Work steadily through the paper.
- Don't spend too long on one question.
- If you have time at the end, go back and check your answers.

Exam Set MEHPI41

International GCSE Mathematics

Formula Sheet for Higher Tier

Arithmetic series

Sum to n terms, $S_n = \frac{n}{2}[2a + (n-1)d]$

Area of trapezium $= \frac{1}{2}(a+b)h$

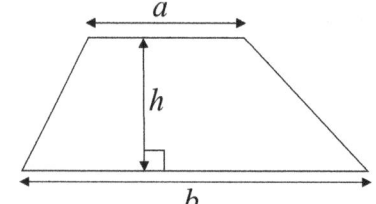

The quadratic equation

The solutions of $ax^2 + bx + c = 0$, where $a \neq 0$, are given by:

$$x = \frac{-b \pm \sqrt{b^2 - 4ac}}{2a}$$

Curved surface area of cone $= \pi r l$

Volume of cone $= \frac{1}{3}\pi r^2 h$

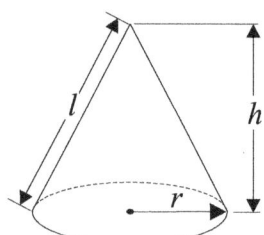

For any triangle ABC:

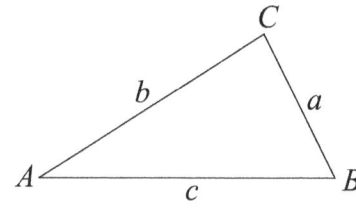

Sine rule: $\frac{a}{\sin A} = \frac{b}{\sin B} = \frac{c}{\sin C}$

Cosine rule: $a^2 = b^2 + c^2 - 2bc \cos A$

Area of triangle $= \frac{1}{2}ab \sin C$

Volume of sphere $= \frac{4}{3}\pi r^3$

Surface area of sphere $= 4\pi r^2$

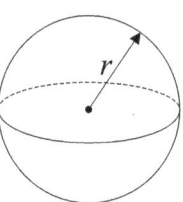

Volume of prism
= area of cross section × length

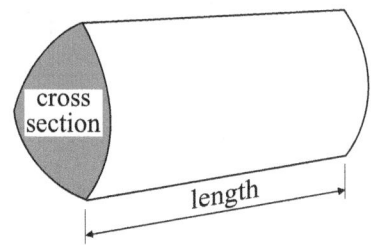

Volume of cylinder $= \pi r^2 h$

Curved surface area of cylinder $= 2\pi r h$

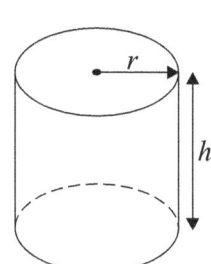

Answer ALL twenty-three questions.

Write your answers in the spaces provided.

You must show all of your working.

1 Use a ruler and compasses to construct the perpendicular bisector of line segment *AB*.
  You must show your construction marks.

  A•─────────────────────────•*B*

  [Total 2 marks]

2 A solid cube with volume 729 cm³ rests on one of its faces on horizontal ground.
  The pressure exerted by the cube on the ground is 1.6 N/cm².

  Find the force exerted by the cube on the ground in newtons (N).

  .... N

  [Total 3 marks]

3 A box contains 123 chocolates, which are either plain, milk or white chocolate.

The ratio of plain to milk chocolates is 2 : 3.
The ratio of milk to white chocolates is 7 : 2.

Find the number of plain chocolates in the box.

....
[Total 3 marks]

4 Show that $3\frac{1}{4} \times 1\frac{3}{5} = 5\frac{1}{5}$

[Total 2 marks]

5 A fair 6-sided dice and a fair 10-sided dice are rolled repeatedly over the course of a game.

The 6-sided dice (with sides numbered 1-6) is rolled 300 times.
The 10-sided dice (with sides numbered 1-10) is rolled 200 times.

Calculate an estimate for the number of times a prime number is rolled.

....
[Total 2 marks]

6 (a) Write the expression $2x^2 - 8x + 5$ in the form $a(x + b)^2 + c$.

....
[2]

(b) Hence, find the exact solutions to $2x^2 - 8x + 5 = 0$.

....
[2]

[Total 4 marks]

7 Calculate $(3.2 \times 10^4) \div (8 \times 10^{-k})$.
Give your answer in standard form in terms of k.

....
[Total 2 marks]

8 The third term of an arithmetic series is 18.
The seventh term is 38.

(a) Write an expression for the nth term of the series.

....
[3]

(b) Work out the sum of the first 50 terms of this series.

....
[2]

[Total 5 marks]

9 For all values of x, the functions f and g are defined by

$$f : x \mapsto 3x - 5$$
$$g : x \mapsto x^2 - 3$$

(a) (i) Find the value of g(−2).

.... [1]

(ii) Write down the range of g.

.... [1]

(b) Find $f^{-1}(x)$.

$f^{-1}(x)$ = ... [2]

(c) Show that $gf(x) = 9x^2 - 30x + 22$

[2]

[Total 6 marks]

10 Rearrange $k + 2 = \sqrt[3]{\dfrac{1}{j + l^2}}$ to make l the subject, where $l > 0$.

....
[Total 4 marks]

11 Show algebraically that the straight line through the points (2, 7) and (5, 13) is perpendicular to the line $2y = 13 - x$.

[Total 3 marks]

12 The diagram shows triangle ABC split into two right-angled triangles.
Angle ABC is 34°.
DB = 6 cm, AC = 5.4 cm

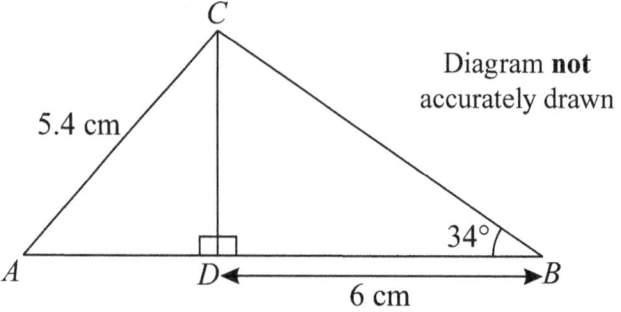

Diagram **not** accurately drawn

Find the area of triangle ABC.
Give your answer correct to 2 decimal places.

.... cm²

[Total 4 marks]

13 The incomplete Venn diagram shows how many Year 11 students study Spanish, French and German. There are 50 Year 11 students in total.

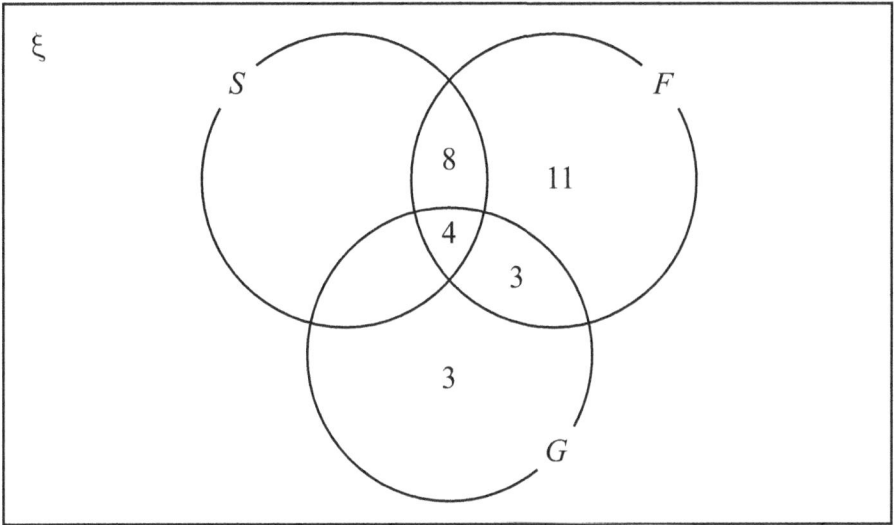

S = {students who study Spanish}, F = {students who study French} and G = {students who study German}.

n(S) = 27 and n($S \cap G$) = 6.

(a) Use this information to complete the Venn diagram.

[2]

Set $A \subset F$ and $A \cap G = \emptyset$

(b) What is the largest possible value of n(A)?

....
[2]

(c) If a student is chosen at random, what is the probability that they study exactly one of the languages?

....
[2]

(d) If a student studying French is chosen at random, what is the probability that they study exactly one of the other languages?

....
[2]

[Total 8 marks]

14 The cumulative frequency graph shows information about the lengths of 60 songs.

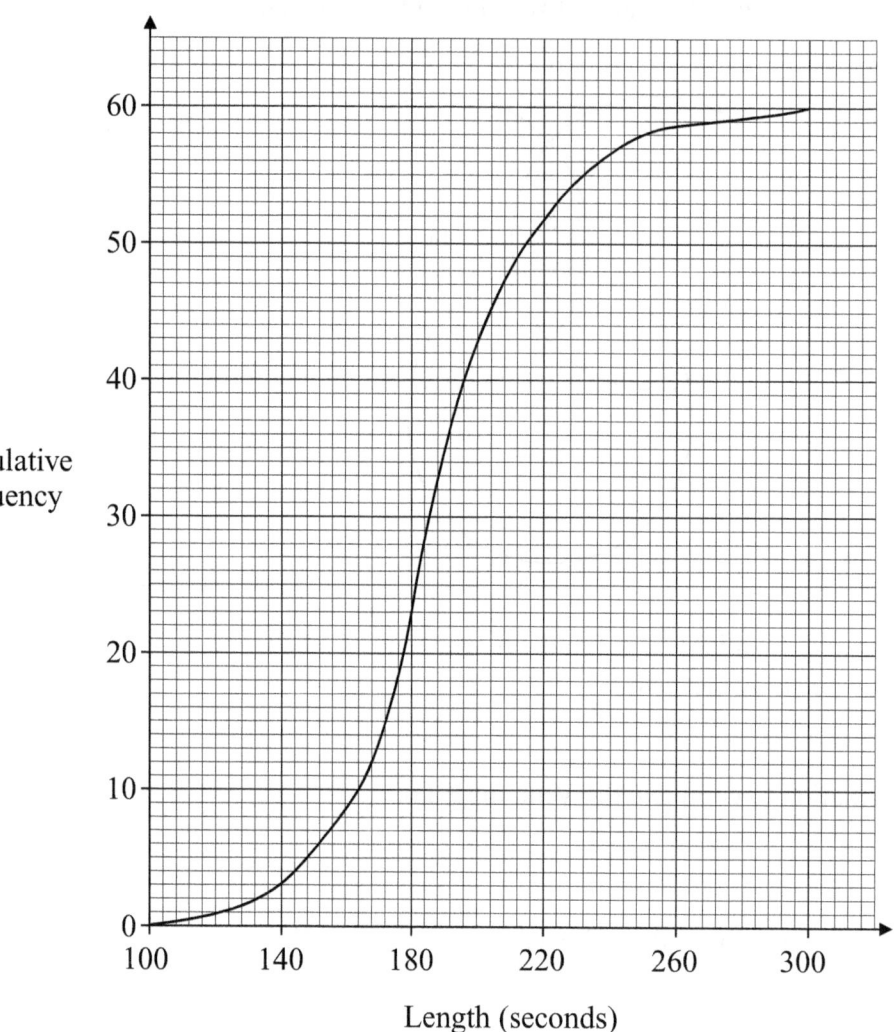

(a) Use the graph to find an estimate for the median length of these songs.

.... seconds
[1]

(b) Use the graph to work out an estimate for the interquartile range.

.... seconds
[2]

[Total 3 marks]

15 The diagram on the right shows a sketch of the graph $y = f(x)$.
The graph passes through the points $(-2, 0)$, $(2, 0)$ and $(0, -3)$.

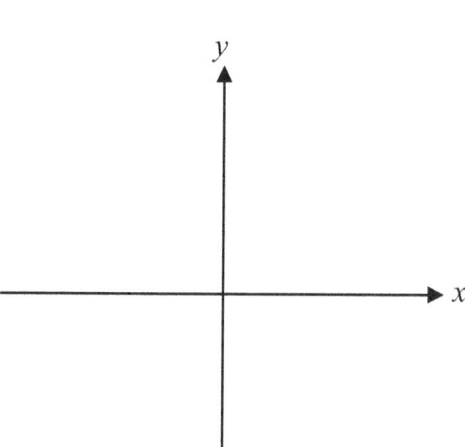

(a) Sketch the graph of $y = f(x + 2)$. Show clearly the points where the graph crosses the x-axis and the y-axis.

[1]

The diagram below shows the graph of a transformation of $f(x)$.

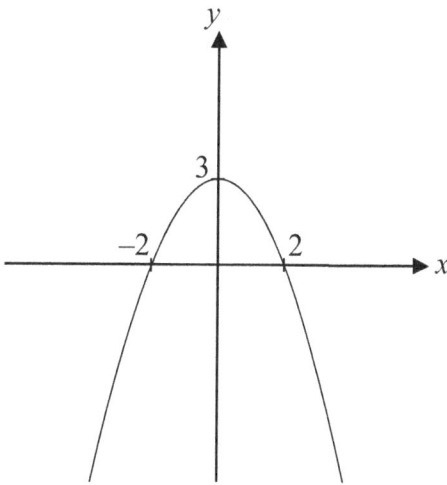

(b) Write down the equation of this curve in terms of f.

....
[1]
[Total 2 marks]

16 The total amount of time employees at Company A spent in meetings in 2019 was 7% greater than in 2018.

In 2019, the number of hours spent in meetings was approximately 32 000.

(a) Estimate how many more hours were spent in meetings in 2019 compared to in 2018. Give your answer correct to 3 significant figures.

.... hours
[3]

The number of hours that employees at Company B spent in meetings in 2019 was 17 000 to the nearest 1000.

To the nearest percentage point, 3% fewer hours were spent in meetings in 2018 than in 2019.

Let n be the total number of hours spent in meetings in 2018 and 2019 by employees at Company B.

(b) Find the upper bound for the value of n.

.... hours
[4]
[Total 7 marks]

17 (a) On the grid below, draw the line that is parallel to $4y + 2x = 3$ and crosses the x-axis at $x = 8$.

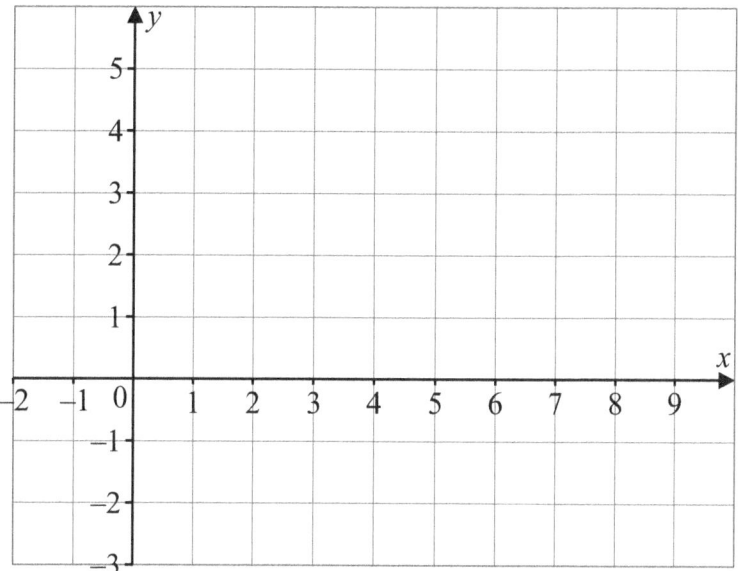

[2]

Let the equation of the line drawn in part (a) be $y = f(x)$, where f is a function of x.

(b) Shade the region that satisfies the following inequalities:

$y \leq f(x)$ $x \geq 1$ $y \geq x - 1$

[3]

[Total 5 marks]

18 The volumes of two spheres are in the ratio 1 : 8.
The surface area of the larger sphere is 28 cm².

What is the radius, r, of the smaller sphere?
Give your answer correct to 2 decimal places.

$r = $ .... cm

[Total 4 marks]

19 At a party, there are *c* children and *a* adults.
The ratio of the number of children to the number of adults at the party is $c:a$.

3 more children and 3 more adults arrive and the ratio is now $2:3$.
Then 2 children leave and 2 more adults arrive, and the ratio becomes $1:2$.

Find the ratio $c:a$ in its lowest terms.

After first change: $\dfrac{c+3}{a+3} = \dfrac{2}{3}$ → $3c - 2a = -3$

After second change: $\dfrac{c+1}{a+5} = \dfrac{1}{2}$ → $2c - a = 3$, so $a = 2c - 3$

Substituting: $3c - 2(2c-3) = -3$ → $-c + 6 = -3$ → $c = 9$, $a = 15$

$c : a = 9 : 15 = 3 : 5$

[Total 6 marks]

20 *A*, *B*, *C* and *D* are points on the circumference of a circle with centre *O*.
Line *EAF* is a tangent to the circle.

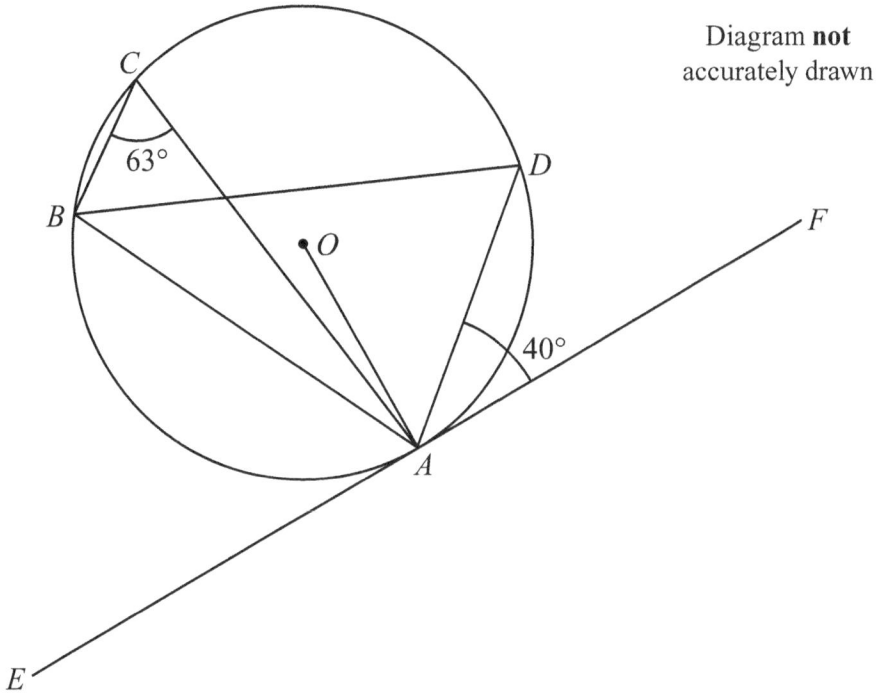

Diagram **not** accurately drawn

Angle *ACB* is 63° and angle *DAF* is 40°.

Find the size of angle *BAO*.
You must give a reason for each stage in your working.

.... °

[Total 6 marks]

21 Find the values of x for which the graph of $y = -\frac{5}{3}x^3 - x^2 + 3x - 6$ has a positive gradient.

....

[Total 6 marks]

22 ABCD is a trapezium.
ABEF is a parallelogram.

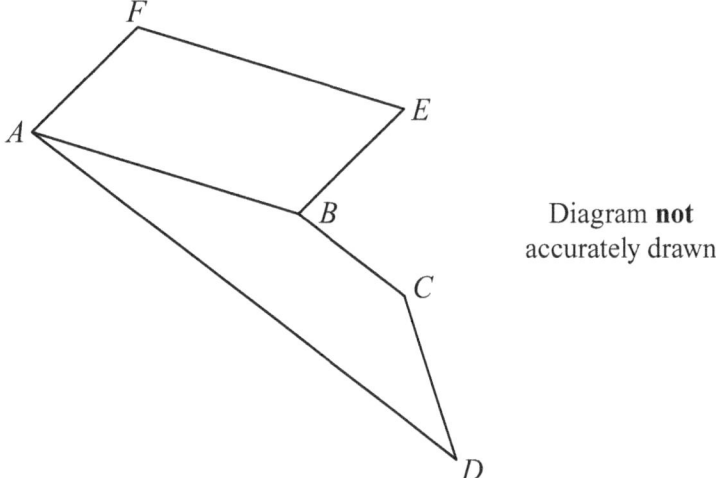

Diagram **not** accurately drawn

$\vec{AD} = \begin{pmatrix} 16 \\ -12 \end{pmatrix}$ $\vec{BF} = \begin{pmatrix} -6 \\ 6 \end{pmatrix}$ $\vec{CE} = \begin{pmatrix} 0 \\ 6 \end{pmatrix}$ $\vec{BC} = 0.25\,\vec{AD}$

Work out the exact magnitude of $\vec{AB}$.

[Total 5 marks]

23 The diagram shows a shape made from a cuboid and a triangular prism.

$AD = 12$ cm, $CD = 8$ cm, $AE = 6$ cm, $EJ = 5$ cm and angle $EJH = 44°$.

(a) Find the size of angle AFC. Give your answer correct to 1 decimal place.

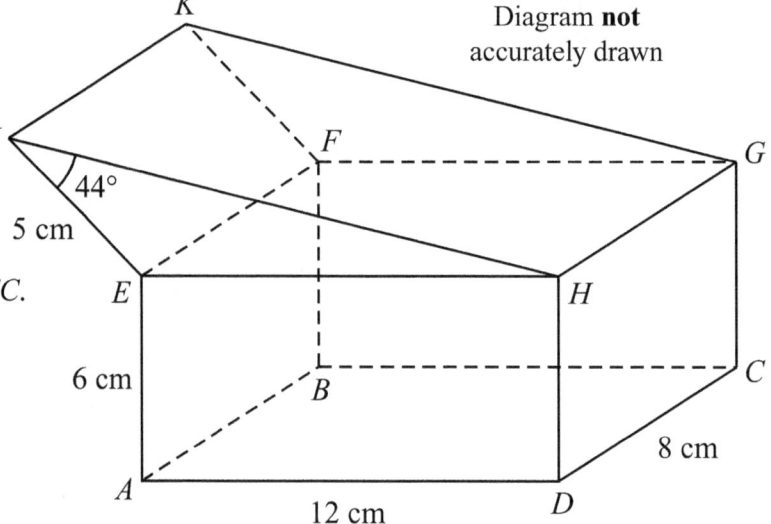

Diagram **not** accurately drawn

.... °
[5]

(b) Find the size of angle EHJ. Give your answer correct to 1 decimal place.

.... °
[3]

[Total 8 marks]

[TOTAL FOR PAPER = 100 MARKS]

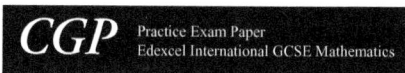

Edexcel International GCSE

Mathematics
Specification A
Higher Tier

**Practice Set 2
Paper 2**

Time allowed: 2 hours

Centre name	
Centre number	
Candidate number	

Surname	
Other names	
Candidate signature	

In addition to this paper you should have:
- A pen, pencil and eraser.
- A calculator.
- A ruler.
- A protractor.
- A pair of compasses.

Tracing paper may be used.

For examiner's use			
Q	Mark	Q	Mark
1		12	
2		13	
3		14	
4		15	
5		16	
6		17	
7		18	
8		19	
9		20	
10		21	
11		22	
Total			

Instructions to candidates
- Write your name and other details in the spaces provided above.
- Answer all questions in the spaces provided.
- In calculations show clearly how you worked out your answers.
- **You may use a calculator.**

Information for candidates
- There are 100 marks available for this paper.
- The marks available are given in brackets at the end of each question.
- You may get marks for method, even if your answer is incorrect.

Advice to candidates
- Work steadily through the paper.
- Don't spend too long on one question.
- If you have time at the end, go back and check your answers.

International GCSE Mathematics

Formula Sheet for Higher Tier

Arithmetic series

Sum to n terms, $S_n = \frac{n}{2}[2a + (n-1)d]$

Area of trapezium $= \frac{1}{2}(a+b)h$

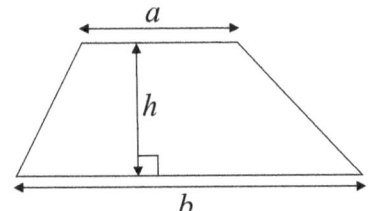

The quadratic equation

The solutions of $ax^2 + bx + c = 0$, where $a \neq 0$, are given by:

$$x = \frac{-b \pm \sqrt{b^2 - 4ac}}{2a}$$

Curved surface area of cone $= \pi r l$

Volume of cone $= \frac{1}{3}\pi r^2 h$

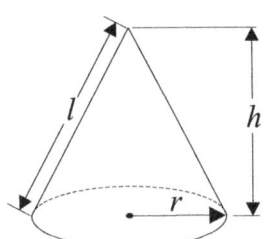

For any triangle ABC:

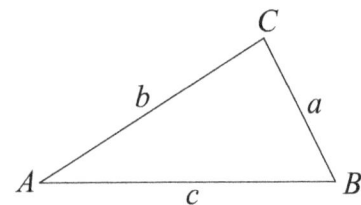

Sine rule: $\quad \dfrac{a}{\sin A} = \dfrac{b}{\sin B} = \dfrac{c}{\sin C}$

Cosine rule: $\quad a^2 = b^2 + c^2 - 2bc\cos A$

Area of triangle $= \frac{1}{2}ab\sin C$

Volume of sphere $= \frac{4}{3}\pi r^3$

Surface area of sphere $= 4\pi r^2$

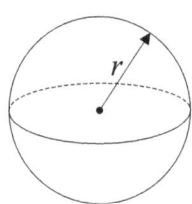

Volume of prism
= area of cross section × length

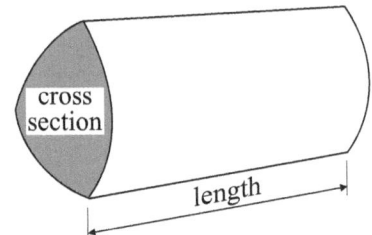

Volume of cylinder $= \pi r^2 h$

Curved surface area of cylinder $= 2\pi r h$

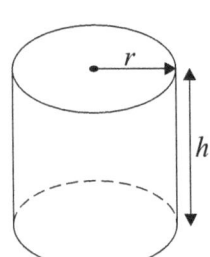

Answer ALL twenty-two questions.

Write your answers in the spaces provided.

You must show all of your working.

1   The mass of a snowflake is 2.5 mg.
    1 000 000 mg = 1 kg.

    Convert the mass of the snowflake into kg.
    Give your answer in standard form.

    .... kg
    [Total 2 marks]

2   (a) Fully factorise $20x^3y + 8xy^2$

    ....
    [1]

    (b) Solve $\dfrac{3x + 4}{2} = \dfrac{5x + 3}{3}$

    $x =$ ....
    [3]

    (c) Solve $2x - 5 < 5x + 4$

    ....
    [2]
    [Total 6 marks]

3 (a) On the grid, enlarge shape **T** by a scale factor $\frac{1}{2}$ with centre of enlargement (5, 7). Label the enlarged shape **S**.

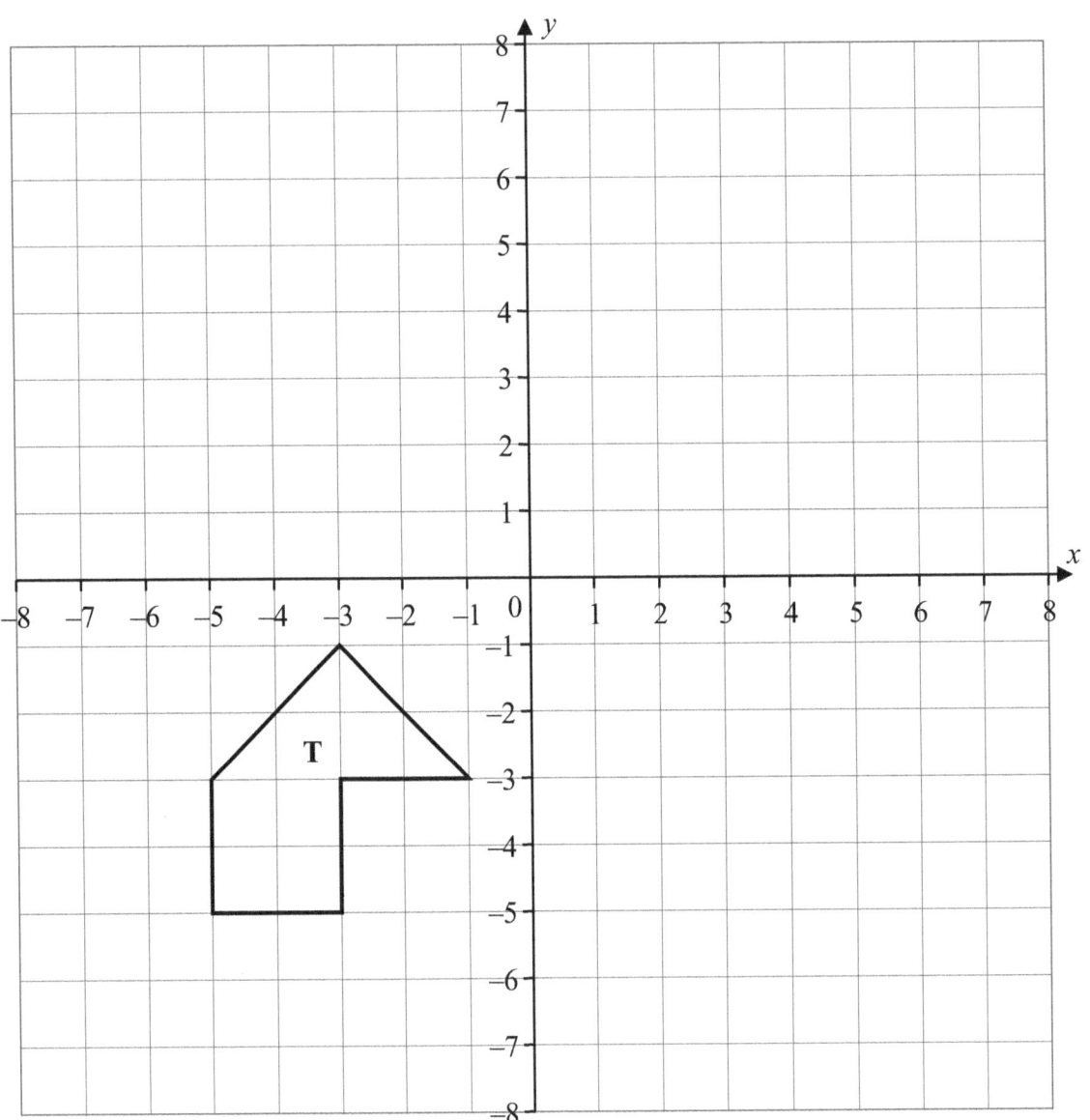

[2]

(b) On the same grid, translate shape **T** by the vector $\begin{pmatrix} 8 \\ -2 \end{pmatrix}$, then translate this shape by the vector $\begin{pmatrix} -2 \\ 3 \end{pmatrix}$.

Label the final translated shape **U**.

[2]

[Total 4 marks]

4 Written as a product of powers of its prime factors, $A = 2 \times 3^2 \times 7$.

(a) Write $B = 392$ as a product of powers of its prime factors.

B = ...
[3]

(b) Find the highest common factor of A and B.

....
[1]

k is the smallest integer that can be multiplied by A to give a square number.

(c) Find the value of k.

k = ...
[1]

[Total 5 marks]

5 The frequency table and histogram show the times taken by entrants to complete a fun run.

Time (*m*) in minutes	Frequency
$10 \leq m < 12$	8
$12 \leq m < 15$	
$15 \leq m < 20$	30
$20 \leq m < 25$	
$25 \leq m < 40$	30

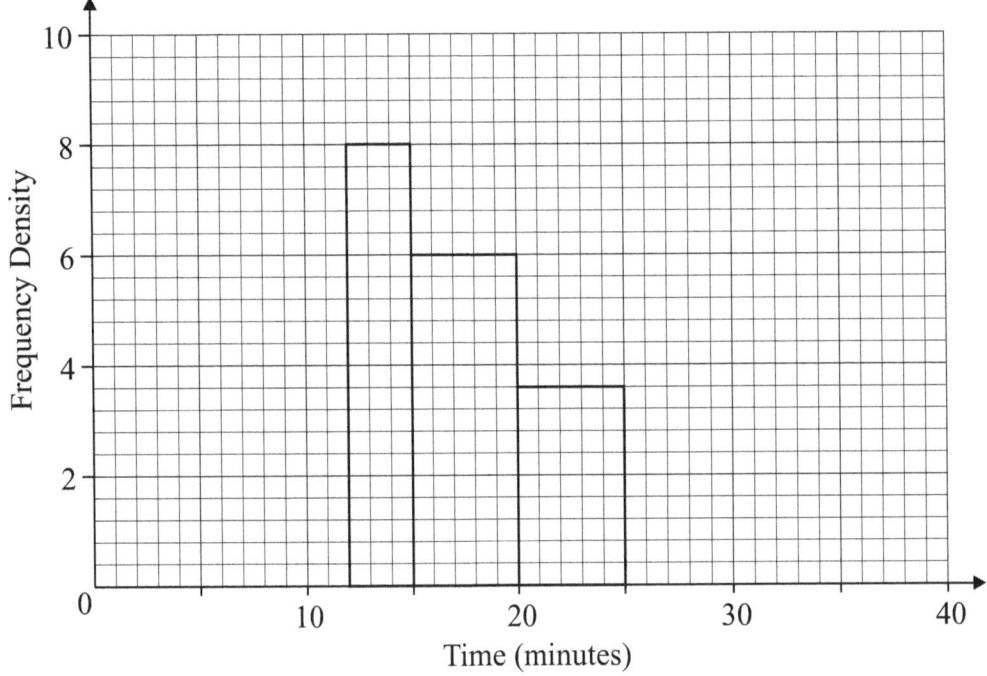

Complete the histogram and frequency table.

[Total 4 marks]

6 The diagram shows part of a 16-sided regular polygon.

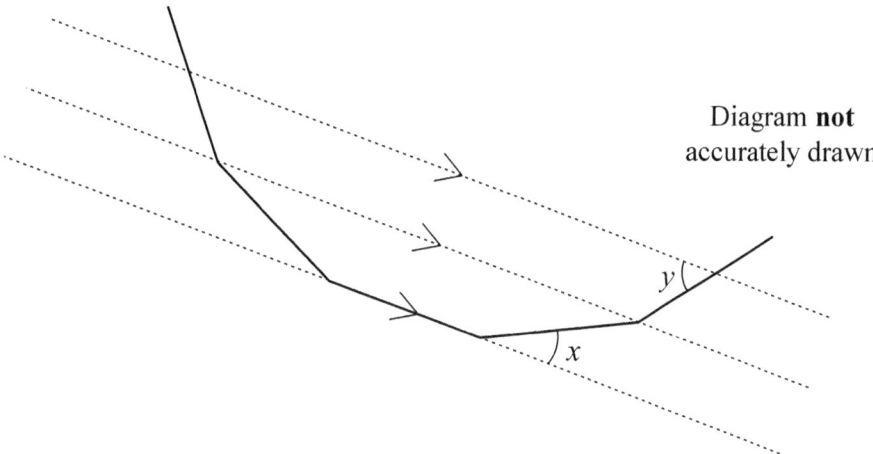

Diagram **not** accurately drawn

(a) Calculate the size of angle *x*.

x = ... °
[1]

(b) Calculate the size of angle *y*.

y = ... °
[2]

[Total 3 marks]

7 Expand and simplify $(2u + 5)(u + 1)(u - 10)$

....
[Total 3 marks]

8 $\mathbf{a} = \begin{pmatrix} 2 \\ -1 \end{pmatrix}$ $\quad \mathbf{b} = \begin{pmatrix} 5 \\ 3 \end{pmatrix}$

If $m\mathbf{a} + n\mathbf{b} = \begin{pmatrix} 1 \\ -6 \end{pmatrix}$, find the values of m and n.

$m = $ ....

$n = $ ....
[Total 4 marks]

9 In a basketball game, Jack takes two shots.

The probability that Jack scores with the first shot is 0.7.
The probability of Jack scoring with both shots is 0.56.
The probability of Jack missing with both shots is 0.18.

(a) Complete the tree diagram to show these probabilities:

First shot   **Second shot**

```
                    ........... → Score
          Score <
  0.7 /         ........... → Miss
     <
  ..... \      ........... → Score
          Miss <
                    ........... → Miss
```

[3]

(b) Given that Jack scores with exactly one shot, is it more likely that Jack missed his first or his second shot? Show your working.

....

[3]

[Total 6 marks]

10 Express $\dfrac{\sqrt{32} + \sqrt{18}}{\sqrt{2} - 1}$ in the form $a + b\sqrt{2}$, where a and b are integers.

....

[Total 4 marks]

11 A, B, C and D are points on the circumference of a circle.

Angle $BAD = x + 18°$
Angle $BCD = 5x + 12°$
Angle $ADC = 43°$

Show that $ABCD$ is a trapezium.

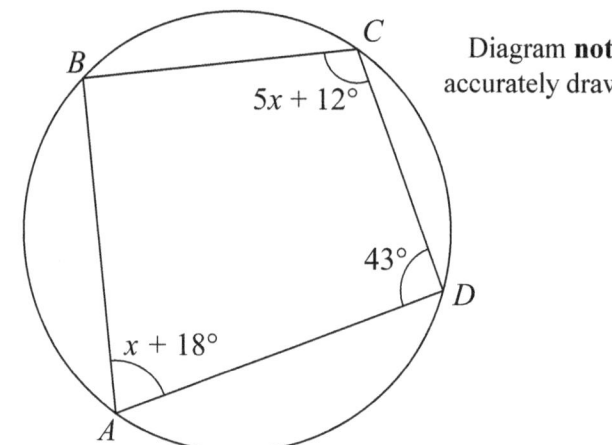

Diagram **not** accurately drawn

[Total 4 marks]

12 c is directly proportional to the square of a. a is always non-negative.
The relationship between c and a is shown in the graph.

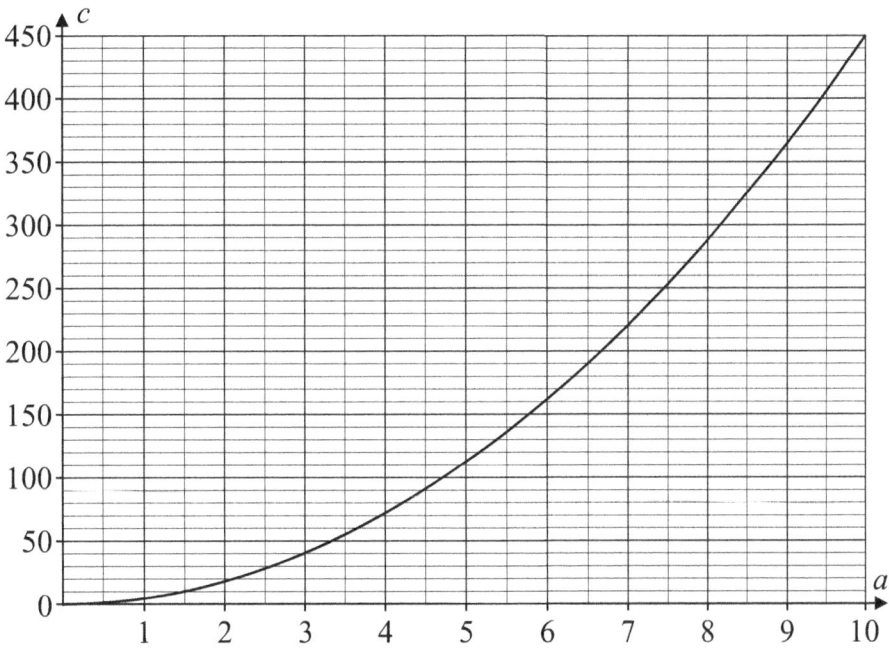

Find the value of a when $c = 882$.

$a = $ ....

[Total 4 marks]

13. $y = -x^3 + 2x^2 + 2x - 1$

(a) Complete the table of values.

x	−1	−0.5	0	0.5	1	1.5	2	2.5
y		−1.375		0.375		3.125		0.875

[2]

(b) Draw the graph of $y = -x^3 + 2x^2 + 2x - 1$ for values of x from −1 to 2.5 on this grid.

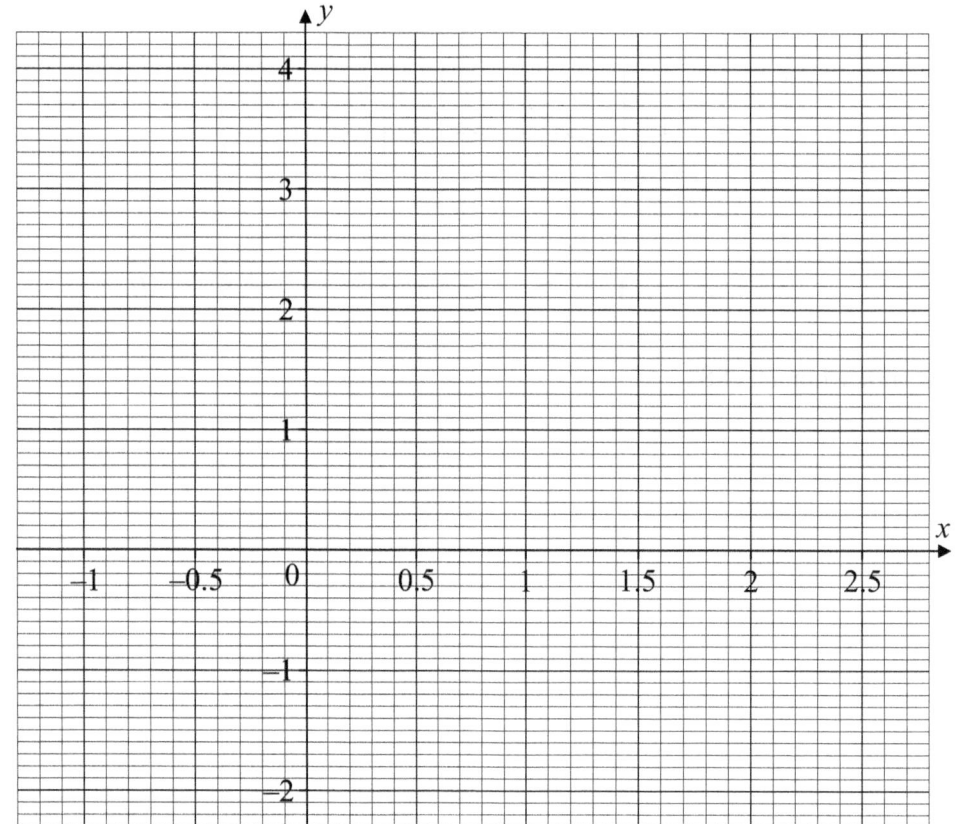

[2]

(c) By drawing a suitable straight line on the grid above, work out estimates of the solutions to the equation

$$-x^3 + 2x^2 = 1$$

Give your answers correct to 1 decimal place.

$x =$ .... , $x =$ .... and $x =$ ....

[4]

[Total 8 marks]

14. The diagram below shows a right-angled triangle.

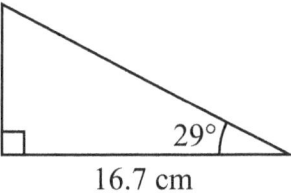

Diagram **not** accurately drawn

A shape is made using two of these triangles and a third triangle.
The three triangles form a right angle at the point where they meet.

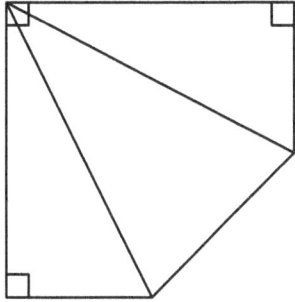

Find the perimeter of the shape.
Give your answer correct to 1 decimal place.

.... cm

[Total 5 marks]

15 The shape below is a sector of a circle.

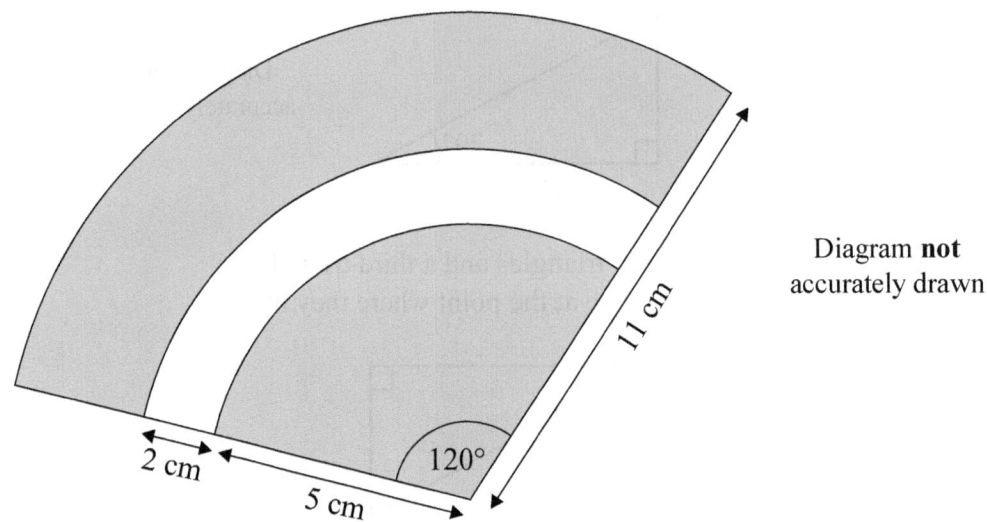

Diagram **not** accurately drawn

Calculate the shaded area of the shape.
Give your answer correct to 3 decimal places.

.... cm²

[Total 4 marks]

16 Fully simplify $\dfrac{6x-3}{2x^2+7x-4} \div \dfrac{15}{x^2-16}$

....
[Total 4 marks]

17 A frustum is made by removing a small cone from the pointed end of a full cone. The two cones are mathematically similar.

The small cone and the frustum have the same height. Find the following ratio in its simplest form.

Volume of the small cone : Volume of frustum

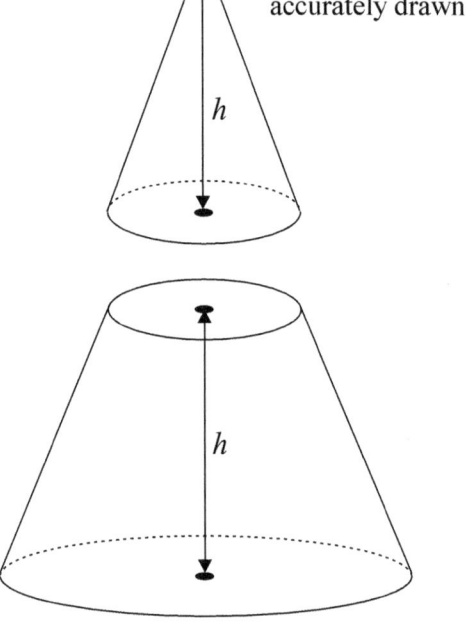

Diagram **not** accurately drawn

[Total 4 marks]

18 Katie is conducting an experiment.

She drops a rubber ball from a height of 140 cm onto concrete.
After each bounce, the ball rises to a height 24% less than the height it fell from.

(a) What height would the ball bounce to after it has hit the ground for the 5th time?
Give your answer correct to 1 decimal place.

.... cm
[3]

Katie tries the experiment three times.

On the second try, she drops the ball from a height $P\%$ less than on the first try.

On the third try, she drops the ball from a height of 178.5 cm.
This is 70% greater than on the second try.

(b) Calculate the value of P.

$P =$ ....
[4]

[Total 7 marks]

19 Christian has a bag containing eleven cards, each labelled with a prime number.
Four of these cards are labelled 2.

Christian picks two cards from the bag at random.

(a) Work out the probability that the product of the numbers on the two cards is odd.

....
[3]

Christian returns the cards to the bag.
He adds a card labelled 0.

He picks two cards from the bag at random.
This time, he replaces each card before picking the next one.

(b) Work out the probability that the product of the numbers on the two cards is 0.

....
[3]
[Total 6 marks]

20 Circle c has its centre at the origin $(0, 0)$.
Line l is a tangent to the circle at the point $P(-8, -6)$.

Find the equation of line l.

....
[Total 4 marks]

21 A particle travels in a straight line.

The velocity, v m/s, of the particle at time t seconds ($t \geq 0$) can be modelled by the equation

$$v = t^3 + 2wt^2 + w^2t + 0.8$$

After exactly 2 seconds, the acceleration of the particle is 10 m/s².

Show that $w = -4 \pm \sqrt{14}$

[Total 4 marks]

22 A rectangular field *ABCD* has lengths such that:

$AB = CD$, $BC = AD$ and $AB > BC$

$AB = \frac{1}{3x}$ miles and $BC = \frac{x}{6}$ miles.

The perimeter of the field is exactly 1 mile.

Find the lengths of sides *AB* and *BC*.

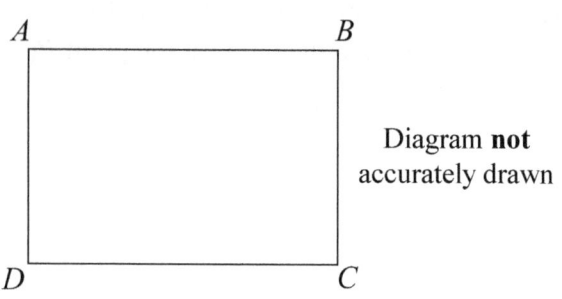

Diagram **not** accurately drawn

AB = ... miles

BC = ... miles

[Total 5 marks]

[TOTAL FOR PAPER = 100 MARKS]

CGP

Edexcel International GCSE

Mathematics

For the Grade 9-1 Course

Practice Exam Papers
Instructions & Answer Book

Higher Tier

Exam Set MEHPI41

Perfect exam practice from CGP!

You can't bluff your way through Edexcel's International GCSE Maths exams. No chance. What you need is a way of making sure you're 100% prepared.

That's where this brilliant pack from CGP comes in. It contains two full sets of realistic mock exams, so you get used to tackling the types of questions examiners love to ask — all in the comfort of your own home / classroom / private jet.

We've also included full answers and mark schemes for each paper, so it's easy to check how you're getting on. You'll be ready for anything when the real exams roll around.

The Three Big Ways to Improve Your Score

1) **Try all of these practice papers**
   These practice papers contain questions in the same style, at the same level and covering the same topics as the questions you could get in the real exam. The more practice you get, the less chance of a nasty surprise on your exam.

2) **Keep practising the things you get wrong**
   The whole point of a practice exam is to find out what you don't know*. So every time you get a question wrong, revise that subject then have another crack at it.

   *Use the mark scheme in this booklet to help you see where you dropped your marks.

3) **Don't throw away easy marks**
   Always answer the question the way it's asked — if it asks for units, use the right ones. Always double-check your answer and don't make silly mistakes — obvious really.

Marking Your Papers

- Do a complete exam (Paper 1 and Paper 2).
- Use the answers and mark scheme in this booklet to mark each exam paper.
- Write down your mark for each paper in the table below — each paper is worth a maximum of 100 marks.
- Find your total for the whole exam (out of a maximum of 200 marks) by adding up your marks from both papers.
- Follow the instructions below to estimate your grade.

	Paper 1	Paper 2	Total	Grade
SET 1				
SET 2				

Estimating Your Grade

- If you want to get a **rough idea** of the grade you're working at, we suggest you compare the **total mark** you got in **each set** to the latest set of grade boundaries.

- Grade boundaries are set for each individual exam, so they're likely to **change** from year to year. You can find the latest set of grade boundaries by going to **www.cgpbooks.co.uk/gcsegradeboundaries**

- Jot down the marks required for each grade in the table below so you don't have to refer back to the website. Use these marks to **estimate your grade**. If you're borderline, don't push yourself up a grade — the real examiners won't.

Total mark required for each grade						
Grade	9	8	7	6	5	4
Total mark out of 200						

- Remember, this will only be a **rough guide**, and grade boundaries will be different for different exams, but it should help you to see how you're getting on.

Published by CGP

Editors: Liam Dyer, Sean McParland, Tom Miles, Caley Simpson.
Contributors: Mark Moody, Kieran Wardell.
Proofreaders: Mona Allen, Glenn Rogers.

Many thanks to Jan Greenway for the copyright research.

Clipart from Corel®

Printed by Elanders Ltd, Newcastle upon Tyne.

Text, design, layout and original illustrations
© Coordination Group Publications Ltd. (CGP) 2020
All rights reserved.

Photocopying this book is not permitted, except where you hold and follow the terms of an appropriate licence.
Extra copies are available from CGP • www.cgpbooks.co.uk

Answers

Set 1 Paper 1

1. The difference between each term is 7, so the expression for the nth term will include a $7n$. *[1 mark]*
   The sequence $7n$ is: 7, 14, 21, 28...
   Each term in the given arithmetic sequence is 4 less than this sequence, so the nth term = $7n - 4$ *[1 mark]*
   [2 marks available in total — as above]

2. a) $\frac{12x^4y^3}{2x^3y^7} = \frac{6x}{y^4}$ or $6xy^{-4}$
   [2 marks available — 1 mark for correct index for x or y, 1 mark for the correct answer]
   b) $3(x - 7) = 7x + 13$
   $3x - 21 = 7x + 13$ *[1 mark]*
   $-34 = 4x$ *[1 mark]*
   $x = -\frac{17}{2}$ or -8.5 *[1 mark]*
   [3 marks available in total — as above]

3. Let $x = 0.411111...$
   So, $10x = 4.11111...$ and $100x = 41.11111...$ *[1 mark]*
   $100x - 10x = 41.11111... - 4.11111...$
   $90x = 37$
   $x = \frac{37}{90}$ *[1 mark]*
   [2 marks available in total — as above]

4. Take the midpoints of each class, multiply by the frequency and divide by the total.
   $(45 \times 27) + (55 \times 30) + (65 \times 16) + (75 \times 7) = 4430$
   $27 + 30 + 16 + 7 = 80$
   So, the mean mass is $4430 \div 80 = 55.375$ g
   [3 marks available — 1 mark for multiplying each midpoint by the corresponding frequency, 1 mark for dividing the sum of the multiplied frequencies by the number of eggs, 1 mark for the correct answer]

5. $16^{-\frac{3}{2}} = \frac{1}{16^{\frac{3}{2}}} = \frac{1}{\sqrt{16}^3} = \frac{1}{4^3} = \frac{1}{64}$
   [2 marks available — 1 mark for either the square root or the reciprocal, 1 mark for a complete correct solution]

6. a) This is $y = \sin x$ stretched vertically by a scale factor of 3, so the maximum point is $(90°, 3)$. *[1 mark]*
   b) This is $y = \sin x$ stretched horizontally by a scale factor of $\frac{1}{2}$, so the maximum point is $(45°, 1)$. *[1 mark]*
   c) This is $y = \sin x$ translated vertically up by 25, so the maximum point is $(90°, 26)$. *[1 mark]*

7. Sketch a right-angled triangle to help. The angle of depression is 32°, so the angle in the triangle at Y is $90° - 32° = 58°$.

   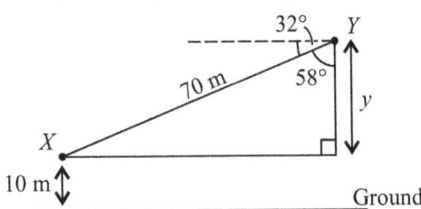

   $\cos 58° = \frac{y}{70}$ *[1 mark]*, so $y = 70 \cos 58°$ *[1 mark]*
   $= 37.094...$ *[1 mark]*
   The height of Y above the ground is
   $y + 10 = 47.1$ m (1 d.p.) *[1 mark]*
   [4 marks available in total — as above]
   Another method would be to use alternate angles: the angle in the triangle at X is 32°. You could then find y using sin 32°.

8. $\frac{2}{7} - \frac{x+1}{x-3} = \frac{2(x-3)}{7(x-3)} - \frac{7(x+1)}{7(x-3)}$
   $= \frac{2x - 6 - 7x - 7}{7(x-3)}$
   $= \frac{-5x - 13}{7(x-3)}$
   [3 marks available — 1 mark for a correct common denominator, 1 mark for a correct single fraction with the brackets expanded, 1 mark for the correct, simplified answer]

9. a)
   ```
   ξ
        M           F
              3
         9    6    2
              12        4
                   5
                10
              8    11    7
   ```
   [3 marks available — 1 mark for the correct values in the circles, 1 mark for the correct values in the overlap, 1 mark for correct values outside of the circles]
   b) $M \cap F = \{3, 6, 12\}$. There are 12 numbers in total, so $P(M \cap F) = \frac{3}{12}$ (or $\frac{1}{4}$). *[1 mark]*
   $(M \cup F)' = \{7, 8, 11\}$ so $P(M \cup F)' = \frac{3}{12}$ (or $\frac{1}{4}$). *[1 mark]*
   [2 marks available in total — as above]
   If you've stated that the probabilities are equal as the set $M \cap F$ has the same number of elements as the set $(M \cup F)'$ you'll still get the marks.

10. a) His new pay is 106% of his pay before the rise, so his pay before the rise was $\frac{€34\,450}{106} \times 100 = €32\,500$.
    [3 marks available — 1 mark for using 106 or 1.06, 1 mark for dividing €34 450 by 106 or 1.06, 1 mark for the correct answer]
    b) Amount after 4 years = $€7300 \times (1 + 0.025)^4$
    $= €8057.83... = €8058$ (nearest €)
    [3 marks available — 1 mark for using 1.025 or 0.025, 1 mark for a complete method to deal with compound interest, 1 mark for the correct answer]
    Alternatively, you could work out the amount in the account after each year by multiplying by 1.025 four times.

11. Find the prime factors:
    $450 = 2 \times 3^2 \times 5^2$, so $450^3 = (2 \times 3^2 \times 5^2)^3 = 2^3 \times 3^6 \times 5^6$
    $240 = 2^4 \times 3 \times 5$, so $240^3 = (2^4 \times 3 \times 5)^3 = 2^{12} \times 3^3 \times 5^3$
    Multiply the prime factors that appear in either number:
    LCM = $2^{12} \times 3^6 \times 5^6 = (2^2 \times 3 \times 5)^6 = 60^6$
    [4 marks available — 1 mark for finding prime factors of 450, 1 mark for finding prime factors of 240, 1 mark for finding the LCM of 450^3 and 240^3, 1 mark for showing the LCM = 60^6]
    You can use factor trees to find the prime factors of each number.

12. a) gradient = $\frac{9-3}{2-0} = \frac{6}{2} = 3$ *[1 mark]*
    You could use different points on L_1 in your working, but your answer should be the same.
    b) Since L_2 is parallel to L_1, its gradient is 3.
    $y - (-7) = 3(x - (-2))$
    $y + 7 = 3x + 6$, so $y = 3x - 1$
    [2 marks available — 1 mark for the correct gradient, 1 mark for the correct equation in the required form]

13 a)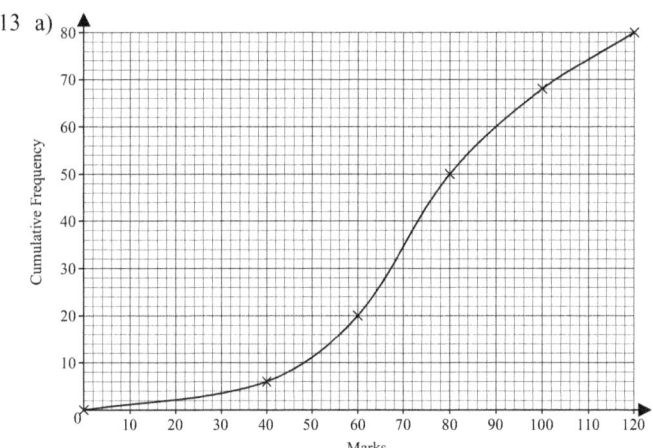

[2 marks available — 1 mark for correctly plotting each point, 1 mark for joining points with a smooth curve or with straight line segments]

You'll still get the marks if you don't join your graph to the origin, but it's a good idea to do so anyway.

b) From the graph, 90 marks gives a cumulative frequency of 60, so there are 80 – 60 = 20 students with a mark of 90 or above.
The ratio 1 : 1.5 = 2 : 3, divides 20 into 20 ÷ 5 × 2 = 8 and 20 ÷ 5 × 3 = 12. So the top 8 students are awarded a platinum certificate.
The lowest of these is the 73rd person which the graph estimates got 108 marks.

[3 marks available — 1 mark for finding number of students with more than 90 marks, 1 mark for finding number of platinum certificates, 1 mark for an answer in the range of 106-110]

You might get slightly different values in your working, depending on the accuracy of your graph. But your final answer should be within the specified range.

14 Consider the upper and lower bound for each measurement by adding or subtracting half of the rounding unit.
400 m: upper bound = 400 + 0.5 = 400.5 m
lower bound = 400 – 0.5 = 399.5 m *[1 mark]*
64.5 s: upper bound = 64.5 + 0.25 = 64.75 s *[1 mark]*
lower bound = 64.5 – 0.25 = 64.25 s

Speed = $\frac{\text{distance}}{\text{time}}$

To get a lower bound on speed, calculate $\frac{\text{lower bound distance}}{\text{upper bound time}}$
= $\frac{399.5}{64.75}$ = 6.1698... = 6.17 m/s (3 s.f.) *[1 mark]*

[3 marks available in total — as above]

15 $-x + 2y = 6 \xrightarrow{\times 3} -3x + 6y = 18$

$\begin{aligned} -3x + 6y &= 18 \\ - \quad 7x + 6y &= 8 \\ \hline -10x &= 10 \\ x &= -1 \end{aligned}$

$-(-1) + 2y = 6$
$1 + 2y = 6$
$2y = 5$
$y = \frac{5}{2}$ or 2.5

[3 marks available — 1 mark for a correct method to eliminate either x or y, 1 mark for a correct method to find the second variable, 1 mark for the correct values of both x and y]

There are a number of ways to solve the equations. For instance, you could instead have multiplied −x + 2y = 6 by 7, or you could have rearranged one of the equations to make x or y the subject and then substituted this into the other equation.

16 The length of VW is $\frac{8d^2 - 4d}{d}$ = (8d − 4) mm. *[1 mark]*
The tile is a parallelogram, so its area is A = VW × width
512 = (8d − 4)(8d + 12) *[1 mark]*
= 64d² − 32d + 96d − 48
= 64d² + 64d − 48
So 0 = 64d² + 64d − 560 *[1 mark]*
= 4d² + 4d − 35
Use the quadratic formula with a = 4, b = 4 and c = −35:
$d = \frac{-4 \pm \sqrt{4^2 - 4 \times 4 \times (-35)}}{2 \times 4}$ *[1 mark]*
$= \frac{-4 \pm \sqrt{576}}{8} = \frac{-4 \pm 24}{8}$
So $d = -\frac{7}{2}$ or $d = \frac{5}{2}$. If $d = -\frac{7}{2}$, then VW = $8 \times -\frac{7}{2} - 4 = -32$.
But VW is a length, so must be positive. So $d = \frac{5}{2}$. *[1 mark]*
Then VW = $8 \times \frac{5}{2} - 4 = 16$ mm. *[1 mark]*

[6 marks available in total — as above]

17 The surface areas are in the ratio 3.6 : 1,
so the side lengths are in the ratio $\sqrt{3.6}$: 1 *[1 mark]*
So the volumes are in the ratio $(\sqrt{3.6})^3$: 1 = 6.83... : 1 *[1 mark]*
The volume of the smaller cylinder is
265 ÷ 6.83... = 38.79... cm³ *[1 mark]*
The volume of a cylinder is V = πr²h, so the height of the smaller cylinder is $h = \frac{V}{\pi r^2} = \frac{38.79...}{\pi(2.2)^2}$ = 2.55... cm
= 2.6 cm (1 d.p.) *[1 mark]*

[4 marks available in total — as above]

You could instead use the side length ratio to find the radius of the larger cylinder, then use the volume to find the height of the larger cylinder. You can then use the side length ratio again to find the height of the smaller cylinder.

18 AC = 9 × sin 42° *[1 mark]* = 6.0221... cm *[1 mark]*
Calculate angle ADC using the sine rule:
$\frac{\sin ADC}{6.0221...} = \frac{\sin 49°}{13}$ *[1 mark]*
$\sin ADC = \frac{\sin 49° \times 6.0221...}{13}$ = 0.3496...
So, angle ADC = sin⁻¹(0.3496...) = 20.4637...
= 20.5° (3 s.f.) *[1 mark]*

[4 marks available in total — as above]

19 a) (2x + 3)² = 4x² *[1 mark]*
4x² + 12x + 9 = 4x²
12x + 9 = 0 *[1 mark]*
$x = -\frac{9}{12} = -\frac{3}{4}$ (= −0.75) *[1 mark]*

[3 marks available in total — as above]

You could also square-root both sides to get 2x + 3 = ±2x. You'd then need to notice that the right-hand side can only be −2x, as no value of x gives 2x + 3 = 2x.

b) f(x) = 27
2x + 3 = 27
2x = 24 so x = 12 *[1 mark]*
To find f⁻¹(x), set x = 2y + 3, and rearrange to make y the subject.
2y = x − 3
$y = \frac{x - 3}{2}$ = f⁻¹(x) = g(x) *[1 mark]*

g(12) = $\frac{12 - 3}{2}$ = 4.5

gg(12) = $\frac{4.5 - 3}{2}$ = 0.75 *[1 mark]*

[3 marks available in total — as above]

c) For h(x) to be defined, f(x) must be positive (you can't take the square root of a negative number and you can't divide by 0). So values of x that satisfy f(x) ≤ 0 cannot be included in the domain of h.
$2x + 3 \le 0$ *[1 mark]*
$2x \le -3$
$x \le -1.5$ *[1 mark]*
[2 marks available in total — as above]

20 If radius of shaded circle = r
Area of shaded circle = πr^2 *[1 mark]*
Radius of sector = $2r$
Area of sector = $\dfrac{360° - a}{360°} \times \pi \times (2r)^2$
$= \dfrac{360° - a}{360°} \times 4\pi r^2$ *[1 mark]*

Area of shaded circle = 30% of area of sector
$\pi r^2 = 0.3 \times \dfrac{360° - a}{360°} \times 4\pi r^2$ *[1 mark]*
$1 = 1.2 \times \dfrac{360° - a}{360°}$
$\dfrac{360°}{1.2} = 360° - a$
$300° = 360° - a$
$a = 60°$ *[1 mark]*
[4 marks available in total — as above]

21 Angle $BCE = n$ = angle ABC (alternate angles are equal) *[1 mark]*
Angle $AOC = 2 \times$ angle $ABC = 2n$ (angle at the centre is twice the angle at the circumference) *[1 mark]*
Angle DAO and angle DCO = 90° (angle where a tangent and a radius meet is 90°) *[1 mark]*
$y = 360° - 90° - 2n - 90°$ (angles in a quadrilateral add up to 360°) = $180° - 2n$ *[1 mark]*
[4 marks available in total — as above]
There are many different ways to complete this question. If you used a correct method and explained your reasons you'll still get the marks.

22 $\overrightarrow{AC} = \overrightarrow{AO} + \overrightarrow{OC} = \overrightarrow{OC} - \overrightarrow{OA} = \mathbf{c} - \mathbf{a}$
$AD:DC = 3:2$ means $\overrightarrow{AD} = \dfrac{3}{5}\overrightarrow{AC}$ *[1 mark]*
So $\overrightarrow{AD} = \dfrac{3}{5}(\mathbf{c} - \mathbf{a}) = \dfrac{3}{5}\mathbf{c} - \dfrac{3}{5}\mathbf{a}$
$\overrightarrow{OD} = \overrightarrow{OA} + \overrightarrow{AD} = \mathbf{a} + \dfrac{3}{5}\mathbf{c} - \dfrac{3}{5}\mathbf{a} = \dfrac{2}{5}\mathbf{a} + \dfrac{3}{5}\mathbf{c}$ *[1 mark]*
E is $\dfrac{2}{3}$ of the way along CB, so
$\overrightarrow{OE} = \overrightarrow{OC} + \dfrac{2}{3}\overrightarrow{CB} = \mathbf{c} + \dfrac{2}{3}\mathbf{a}$ *[1 mark]*
If ODE is a straight line $\overrightarrow{OE}$ must be a multiple of $\overrightarrow{OD}$ so
$\mathbf{c} + \dfrac{2}{3}\mathbf{a} = k(\dfrac{2}{5}\mathbf{a} + \dfrac{3}{5}\mathbf{c})$ for some value of k. *[1 mark]*
Compare coefficients of $\mathbf{c}$ to find k: $\mathbf{c} = \dfrac{3k}{5}\mathbf{c}$ so $k = \dfrac{5}{3}$
$k\overrightarrow{OD} = \dfrac{5}{3}(\dfrac{2}{5}\mathbf{a} + \dfrac{3}{5}\mathbf{c}) = \dfrac{2}{3}\mathbf{a} + \mathbf{c} = \overrightarrow{OE}$ *[1 mark]*
So ODE is a straight line.
[5 marks available in total — as above]
There are many ways to write vectors, so your answer is bound to look a little different to this. Make sure you explain each step and are consistent with how you use your vectors.

23 a) P(red) = r so P(blue) = $1 - r$
P(exactly 1 red) = P(red, blue) + P(blue, red)
$= r \times (1 - r) + (1 - r) \times r = 2r(1 - r)$ *[1 mark]*
So $2r(1 - r) = \dfrac{4}{9}$
$2r - 2r^2 = \dfrac{4}{9}$
$18r - 18r^2 = 4$
$18r^2 - 18r + 4 = 0$ *[1 mark]*
$9r^2 - 9r + 2 = 0$
$(3r - 1)(3r - 2) = 0$ *[1 mark]*
$r = \dfrac{1}{3}$ or $r = \dfrac{2}{3}$ *[1 mark]*
[4 marks available in total — as above]
If you're struggling to make an equation for r, you could use a tree diagram to help you find P(exactly 1 red).

b) The ratio of red beads to blue beads is $\dfrac{1}{3}:\dfrac{2}{3}$ (or $\dfrac{2}{3}:\dfrac{1}{3}$)
= 1 : 2 (or 2 : 1). So there are twice as many of one coloured bead as the other.
Isaac is not correct, as there can't be an odd number of both red beads and blue beads, as odd × 2 = even and even × 2 = even.
[2 marks available — 1 mark for saying that there are twice as many of one coloured bead as the other, 1 mark for a correct conclusion]

24 Differentiate the expression, then set it equal to 0:
$\dfrac{dy}{dx} = 3x^2 + 6x - 9 = 0$
$x^2 + 2x - 3 = 0$
$(x - 1)(x + 3) = 0$
So $x = 1$ and $x = -3$
When $x = 1$, $y = 1^3 + 3(1)^2 - 9(1) + 7$
$= 1 + 3 - 9 + 7 = 2$
When $x = -3$, $y = (-3)^3 + 3(-3)^2 - 9(-3) + 7$
$= -27 + 27 + 27 + 7 = 34$
So the two stationary points are (1, 2) and (–3, 34).
[6 marks available — 1 mark for differentiating one term correctly, 1 mark for a fully correct derivative, 1 mark for setting the derivative equal to zero, 1 mark for a correct method to solve the quadratic, 1 mark for the correct x-coordinates, 1 mark for the correct coordinates of the stationary points]

Set 1 Paper 2

1. $n^2 = (2^3 \times 3^2 \times 5)^2 = 2^6 \times 3^4 \times 5^2$ *[1 mark]*

2. a) $3\mathbf{a} = 3 \times \begin{pmatrix} 8 \\ 3 \end{pmatrix} = \begin{pmatrix} 24 \\ 9 \end{pmatrix}$ *[1 mark]*

   b) $\mathbf{a} - 4\mathbf{b} = \begin{pmatrix} 8 \\ 3 \end{pmatrix} - 4\begin{pmatrix} 1 \\ -7 \end{pmatrix} = \begin{pmatrix} 8 \\ 3 \end{pmatrix} - \begin{pmatrix} 4 \\ -28 \end{pmatrix}$ *[1 mark]*
   $= \begin{pmatrix} 4 \\ 31 \end{pmatrix}$ *[1 mark]*
   [2 marks available in total — as above]

   c) $\mathbf{a} + \mathbf{b} = \begin{pmatrix} 8 \\ 3 \end{pmatrix} + \begin{pmatrix} 1 \\ -7 \end{pmatrix} = \begin{pmatrix} 9 \\ -4 \end{pmatrix}$
   Magnitude $= \sqrt{9^2 + (-4)^2}$ *[1 mark]*
   $= \sqrt{81 + 16} = \sqrt{97}$ *[1 mark]*
   [2 marks available in total — as above]

3. a) $P = \{2, 3, 5, 7, 11, 13, 17, 19\}$ and $Q = \{1, 2, 3, 4, 6, 8, 12\}$
   $P \cap Q$ is the set of numbers that appear in both sets $= \{2, 3\}$
   [2 marks available — 2 marks if both elements are correct, otherwise 1 mark if one element is omitted or if an extra element is included]

   b) $n(P \cup Q)$ is the number of members in the set $P \cup Q$.
   $P \cup Q = \{1, 2, 3, 4, 5, 6, 7, 8, 11, 12, 13, 17, 19\}$ *[1 mark]*
   $n(P \cup Q) = 13$ *[1 mark]*
   [2 marks available in total — as above]

4. £11 367 = 3 shares, so 1 share = £11 367 ÷ 3 = £3789
   Alison's share is £3789 × 7 = £26 523 *[1 mark]*
   Che's share is £3789 × 2 = £7578 *[1 mark]*
   Alison gets £26 523 − £7578 = £18 945 more than Che. *[1 mark]*
   [3 marks available in total — as above]
   There is a difference of 7 − 2 = 5 shares between Alison and Che, so you could do £3789 × 5 = £18 945 too.

5. a) $(4, 0) - (-2, -1) = (6, 1)$ so A translated 6 units to the right and 1 unit upwards gives B.
   As a column vector, this is $\begin{pmatrix} 6 \\ 1 \end{pmatrix}$ *[1 mark]*

   b) C lies in the opposite direction from A than B and AC is half the distance as length AB.
   So the column vector is $-\frac{1}{2} \times \begin{pmatrix} 6 \\ 1 \end{pmatrix} = \begin{pmatrix} -3 \\ -\frac{1}{2} \end{pmatrix}$ *[1 mark]*

6. Put the data in order. Since m is the median, it'll be in the middle:
   26 27 34 <u>35</u> 35 36 37 m 37 38 41 <u>42</u> 43 45 45
   The lower quartile is 35 and the upper quartile is 42.
   Interquartile range = 42 − 35 = 7
   [3 marks available — 1 mark for putting the data in order or identifying one of 35 or 42, 1 mark for identifying both of 35 and 42, 1 mark for the correct answer]

7. First term (a) = number of white dots in first pattern = 4
   Common difference (d) = 3
   Now use the formula for the sum of an arithmetic series:
   $S_{1000} = \frac{1000}{2}(2 \times 4 + (1000 - 1) \times 3) = 1\,502\,500$
   [3 marks available — 1 mark for identifying at least one of the first term or common difference, 1 mark for substituting correctly into the formula for the sum of the sequence, 1 mark for the correct answer]

8. Arc length $= \frac{150°}{360°} \times \pi \times (6 \times 2)$ *[1 mark]*
   $= \frac{5}{12} \times \pi \times 12 = 5\pi$ cm *[1 mark]*
   Perimeter = 6 + 6 + 5π = 27.707... cm = 27.7 cm (3 s.f.) *[1 mark]*
   [3 marks available in total — as above]

9. 5543 pesos is 120.5% of the pesos he originally had.
   Original amount $= \frac{5543}{120.5} \times 100 = 4600$ pesos
   Convert to pounds (£): $\frac{4600}{138} \times 5 = £166.66...$
   $= £167$ (to the nearest pound)
   [4 marks available — 1 mark for using 120.5 or 1.205, 1 mark for dividing 5543 by 120.5 or 1.205, 1 mark for the correct original number of pesos, 1 mark for the correct answer]
   You could also have answered this question by converting 5543 pesos into pounds first, and then working out the original number of pounds.

10. $2y = \frac{3x}{2-5x}$
    $2y(2 - 5x) = 3x$ *[1 mark]*
    $4y - 10xy = 3x$
    $4y = 3x + 10xy$ *[1 mark]*
    $4y = x(3 + 10y)$
    $x = \frac{4y}{3 + 10y}$ *[1 mark]*
    [3 marks available in total — as above]

11. Use the sine rule:
    $\frac{10}{\sin XZY} = \frac{6.8}{\sin 22°}$ *[1 mark]*, so $\sin XZY = \frac{10 \sin 22°}{6.8}$
    $\sin^{-1}\left(\frac{10 \sin 22°}{6.8}\right)$ *[1 mark]* $= 33.428...°$ *[1 mark]*
    Since angle XYZ is acute, this value is not large enough to be the size of angle XZY. (If it were, then the angles in the triangle could add up to no more than 90° + 22° + 33.428...° < 180°.)
    So angle XZY must be the obtuse angle with the same sine value, so subtract this value from 180°:
    Angle $XZY = 180° - 33.428...° = 146.571...°$
    $= 146.57°$ (2 d.p.) *[1 mark]*
    [4 marks available in total — as above]

12. Multiply two brackets out first and then multiply the product by the third bracket.
    $(x + 3)(x + 5) = x^2 + 5x + 3x + 15 = x^2 + 8x + 15$
    $(x^2 + 8x + 15)(x - 2) = x^3 - 2x^2 + 8x^2 - 16x + 15x - 30$
    $= x^3 + 6x^2 - x - 30$
    [3 marks available — 1 mark for multiplying two pairs of brackets correctly, 1 mark for multiplying three pairs of brackets correctly, 1 mark for simplifying correctly]

13. $\min(F) = \min(r) \times \min(s) - 2 \times \max(t)$
    $= 49.5 \times 4.05 - 2 \times 0.5$
    $= 199.475$
    [3 marks available — 1 mark for using 49.5 or 4.05, 1 mark for using 0.5, 1 mark for the correct answer]

14. a) The class with the greatest frequency is represented by the bar with the greatest area.
    $0 < t \leq 10$ is clearly smaller than $20 < t \leq 30$
    $10 < t \leq 15$: 5 × 9.6 = 48 students
    $15 < t \leq 20$ is clearly smaller than $10 < t \leq 15$
    $20 < t \leq 30$: 10 × 2.8 = 28 students
    $30 < t \leq 60$: 30 × 0.5 = 15 students
    So the class with the greatest frequency is $10 < t \leq 15$.
    [2 marks available — 1 mark for a correct method to work out the frequency of at least one class, 1 mark for the correct answer with clear working]

    b) 8:40 am − 8:15 am = 25 minutes *[1 mark]*
    To estimate the number of students who take 25 minutes or more, take half of those in the range $20 < t \leq 30$ plus all of those in the range $30 < t \leq 60$. *[1 mark]*
    $20 < t \leq 30$: 10 × 2.8 = 28 students from part (a)
    $30 < t \leq 60$: 30 × 0.5 = 15 students from part (a)
    $\frac{28}{2} + 15 = 29$ students, so 29 students need to leave before 8:15 am to arrive at school on time. *[1 mark]*
    [3 marks available in total — as above]

15 Volume = $\frac{\pi \times (1.2)^2 \times 4.3}{3}$ = 6.4842... cm³ *[1 mark]*
    Density = $\frac{17.5}{6.4842...}$ *[1 mark]* = 2.70 g/cm³ (3 s.f.) *[1 mark]*
    [3 marks available in total — as above]

16 $\frac{4}{3+\sqrt{5}} + \sqrt{5} = \frac{4 + \sqrt{5}(3+\sqrt{5})}{3+\sqrt{5}}$ *[1 mark]*
    $= \frac{4 + 3\sqrt{5} + 5}{3+\sqrt{5}} = \frac{9 + 3\sqrt{5}}{3+\sqrt{5}}$ *[1 mark]*
    Take 3 out as a common factor in the numerator and cancel down.
    $\frac{3(3+\sqrt{5})}{3+\sqrt{5}} = 3$ *[1 mark]*
    [3 marks available in total — as above]
    Alternatively, you could multiply the top and bottom of the fraction by $3 - \sqrt{5}$ and then add $\sqrt{5}$ once you've simplified the fraction.

17 The mean is 8, so the total must be 4 × 8 = 32.
    $3x + 7 + 13 + \frac{y}{3} = 32$, so $3x + \frac{y}{3} = 32 - 7 - 13 = 12$
    The range is 16, so this is the difference between the largest and smallest number.
    Check to see if 13 is the largest number: $3x$ is the smallest number, so $3x = 13 - 16 = -3$. But $3x + \frac{y}{3} = 12$ means $\frac{y}{3} = 15$, so 13 can't be the largest number.
    So $\frac{y}{3}$ must be the largest number and $\frac{y}{3} - 3x = 16$.
    Subtract the equations $3x + \frac{y}{3} = 12$ and $\frac{y}{3} - 3x = 16$:
    $3x + \frac{y}{3} - (\frac{y}{3} - 3x) = 12 - 16$
    $6x = -4$, so $x = -\frac{2}{3}$
    $3(-\frac{2}{3}) + \frac{y}{3} = 12$, so $-2 + \frac{y}{3} = 12$ and $y = 3 \times 14 = 42$
    [5 marks available — 1 mark for finding the first equation in x and y, 1 mark for finding the second equation in x and y, 1 mark for discounting 13 with justification, 1 mark for the correct value of x, 1 mark for the correct value of y]
    You can also use trial and error (by trying out different values of x) to get to the correct solution, but you must show your working.

18 $y = \frac{k}{\sqrt{x}}$ *[1 mark]*, $12 = \frac{k}{\sqrt{0.09}} = \frac{k}{0.3}$
    $k = 12 \times 0.3 = 3.6$ *[1 mark]*
    When $x = 0.16$, $y = \frac{3.6}{\sqrt{0.16}} = \frac{3.6}{0.4} = 9$ *[1 mark]*
    [3 marks available in total — as above]

19 a) To find which two countries are closest in terms of population, make all numbers the same power of 10.
        Austria: $8.96 \times 10^6 = 0.896 \times 10^7$
        Compare 3.80, 0.896, 3.19 and 3.43
        3.43 – 3.19 = 0.24 is the smallest difference,
        so the countries that have the closest populations are Malaysia and Saudi Arabia.
        [2 marks available — 1 mark for comparing populations, 1 mark for the correct answer]
    b) To find the population density, divide population by area.
        Afghanistan: $(3.80 \times 10^7) \div (6.52 \times 10^5) = 58.2822...$
        Austria: $(8.96 \times 10^6) \div (8.39 \times 10^4) = 106.7938...$
        Malaysia: $(3.19 \times 10^7) \div (3.30 \times 10^5) = 96.6666...$
        Saudi Arabia: $(3.43 \times 10^7) \div (2.15 \times 10^6) = 15.9534...$
        So the country with the greatest population density is Austria.
        [2 marks available — 1 mark for two or more calculations correct, 1 mark for the correct answer]

20 Her two-digit number can be a multiple of 5 in these ways:
    • First spin does not land on 5, then second spin lands on 0 or 5.
    • First spin does land on 5, then second spin lands on 0.
    Using the AND rule, P(first way) = $\frac{8}{9} \times \frac{2}{9} = \frac{16}{81}$
    Using the AND rule, P(second way) = $\frac{1}{9} \times \frac{1}{9} = \frac{1}{81}$
    Using the OR rule, P(multiple of 5) = $\frac{16}{81} + \frac{1}{81} = \frac{17}{81}$
    [3 marks available — 1 mark for the correct probability of the spinner landing on 0 or 5, 1 mark for using conditional probabilities for the second spin, 1 mark for the correct answer]
    You could have used a sample space diagram to answer this question. In any case, you need to notice that the possibilities for the second spin are different depending on whether the first spin lands on 5 or not — there'll be no 5 on the spinner if it does.

21 Exterior angle = 180° – interior angle = 180° – 160° = 20°
    Use the formula: exterior angle = $\frac{360°}{\text{number of sides}}$
    Rearrange to get: number of sides = $\frac{360°}{\text{exterior angle}}$
    $n = \frac{360°}{20°} = 18$ sides *[1 mark]*, so $4n = 4 \times 18 = 72$ sides
    Exterior angle (72 sides) = $\frac{360°}{72°} = 5°$ *[1 mark]*
    Interior angle (72 sides) = 180° – 5° = 175° *[1 mark]*
    [3 marks available in total — as above]

22 a) $\frac{dy}{dx} = (-5 \times 3)x^2 + \frac{1}{2} - (-1)\frac{13}{x^2} = -15x^2 + \frac{1}{2} + \frac{13}{x^2}$
        [2 marks available — 2 marks for the correct derivative, otherwise 1 mark for at least one term differentiated correctly]
    b) Putting $x = 1$ into the answer from part (a), the gradient of the curve at A is $-15(1)^2 + \frac{1}{2} + \frac{13}{1^2} = -\frac{3}{2}$ *[1 mark]*
        So the gradient of the perpendicular line is $\frac{2}{3}$. *[1 mark]*
        Find the y-coordinate of A by putting $x = 1$ into the equation of the curve:
        $y = -5(1)^3 + \frac{1}{2} - \frac{13}{1} = -\frac{35}{2}$ *[1 mark]*
        So the equation of the line is:
        $y - (-\frac{35}{2}) = \frac{2}{3}(x - 1)$ *[1 mark]*
        $y + \frac{35}{2} = \frac{2}{3}x - \frac{2}{3}$
        $6y + 105 = 4x - 4$
        $4x - 6y - 109 = 0$ *[1 mark]*
        [5 marks available in total — as above]

23 a) Rearrange the linear equation and make x the subject.
        $x - 3y = 10$, so $x = 3y + 10$
        Substitute this equation into $x^2 + y^2 = 20$:
        $(3y + 10)^2 + y^2 = 20$ *[1 mark]*
        Multiply out the brackets and simplify.
        $(3y + 10)^2 + y^2 = 20$
        $(3y + 10)(3y + 10) + y^2 = 20$
        $9y^2 + 60y + 100 + y^2 = 20$
        $10y^2 + 60y + 80 = 0$ *[1 mark]*
        Divide through by 10 to get $y^2 + 6y + 8 = 0$
        $(y + 2)(y + 4) = 0$ *[1 mark]*
        $y = -2$ or $y = -4$
        Use $x = 3y + 10$ to find x.
        When $y = -2$, $x = (3 \times -2) + 10 = 4$ *[1 mark]*
        When $y = -4$, $x = (3 \times -4) + 10 = -2$ *[1 mark]*
        [5 marks available in total — as above]
        You could have rearranged the linear equation to make y the subject but it's a lot trickier to get to the solution.
    b) The graphs will have two points of intersection, one at (4, –2) and one at (–2, –4), as the simultaneous equations have two solutions. *[1 mark]*

24 a) $x^2 - 1 \leq 3(x + 3)$
$x^2 - 1 \leq 3x + 9$
$x^2 - 3x - 10 \leq 0$ *[1 mark]*
Factorise to $(x - 5)(x + 2) = 0$.
So, $x = 5$ and $x = -2$ are the critical values. *[1 mark]*
Use a value in between the critical values to check if it is in the range of the inequality. *[1 mark]*
If $x = 0$, $x^2 - 3x - 10 = 0^2 - (3 \times 0) - 10 = -10$
$-10 \leq 0$, so $x = 0$ is a solution which means the inequality is $-2 \leq x \leq 5$ *[1 mark]*
[4 marks available in total — as above]

b) 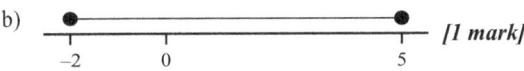 *[1 mark]*

25 a) Let E be the centre of base $ABCD$.
The vertical height of the pyramid is the same as the height of the triangle OAE or OAC.
Work out the length AE by using Pythagoras' theorem.
$AE^2 = 4^2 + 4^2 = 32$
$AE = \sqrt{32}$ cm *[1 mark]*
Work out the height OE using Pythagoras' theorem.
$OE^2 = OA^2 - AE^2 = 12^2 - 32 = 144 - 32 = 112$
$OE = \sqrt{112}$ cm *[1 mark]*
$= \sqrt{16}\sqrt{7} = 4\sqrt{7}$ cm *[1 mark]*
[3 marks available in total — as above]

b) The angle required is angle OAE. Call it x.
$\tan x = \dfrac{OE}{AE} = \dfrac{4\sqrt{7}}{\sqrt{32}}$ *[1 mark]*
$x = \tan^{-1}\left(\dfrac{4\sqrt{7}}{\sqrt{32}}\right)$ *[1 mark]*
$= 61.8744...°$
$= 61.9°$ (3 s.f.) *[1 mark]*
[3 marks available in total — as above]
You could also have used cos or sin to get the correct answer.

26 Use the 'external intersection' property (where two chords are extended to meet outside the circle): $EK \times FK = HK \times GK$
$HK \times GK = (12 + 6) \times 6 = 108$ *[1 mark]*, so $EK \times FK = 108$.
$EK = EJ + JF + FK$
$= FK + 16 + FK$ (since $EJ = FK$)
$= 2FK + 16$
Let $x = FK$. Then the 'external intersection' property gives
$(2x + 16)x = 108$
$2x^2 + 16x = 108$
$2x^2 + 16x - 108 = 0$ *[1 mark]*, so $x^2 + 8x - 54 = 0$
Use the quadratic formula with $a = 1$, $b = 8$ and $c = -54$:
$x = \dfrac{-8 \pm \sqrt{8^2 - (4 \times 1 \times -54)}}{2 \times 1}$ *[1 mark]*
$= 4.36...$ or $-12.36...$
x is a length and must be positive, so $x = FK = 4.36...$ cm *[1 mark]*
So $JK = 16 + 4.36... = 20.36...$ cm
Area of $HJK = \dfrac{1}{2} \times 18 \times 20.36... \times \sin 35°$ *[1 mark]*
$= 105.13...$ cm^2
$= 105$ cm^2 (3 s.f.) *[1 mark]*
[6 marks available in total — as above]

Set 2 Paper 1

1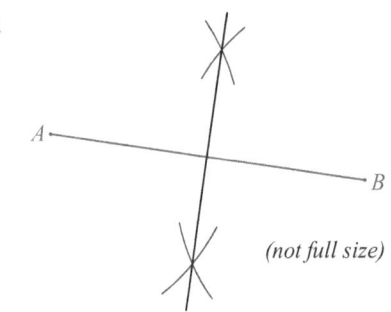

(not full size)

[2 marks available — 1 mark for the correct construction marks, 1 mark for the correctly drawn perpendicular bisector]

2 Volume = 729 cm^3, so side length = $\sqrt[3]{729}$ = 9 cm
So area of a face = 9^2 = 81 cm^2 *[1 mark]*
$1.6 = \dfrac{\text{force}}{81}$ *[1 mark]*, so force = $1.6 \times 81 = 129.6$ N *[1 mark]*
[3 marks available in total — as above]

3 The values in the two ratios that represent 'milk' are 3 and 7. 21 is the lowest common multiple of 3 and 7, so multiply the first ratio by 7 to get 14 : 21, and the second ratio by 3 to get 21 : 6. *[1 mark]*
So, the ratio of all three chocolates is 14 : 21 : 6. *[1 mark]*
There are $14 + 21 + 6 = 41$ parts, so one part is $123 \div 41 = 3$ chocolates. There are $14 \times 3 = 42$ plain chocolates. *[1 mark]*
[3 marks available in total — as above]

4 $3\dfrac{1}{4} = \dfrac{13}{4}$ and $1\dfrac{3}{5} = \dfrac{8}{5}$
$3\dfrac{1}{4} \times 1\dfrac{3}{5} = \dfrac{13}{4} \times \dfrac{8}{5} = \dfrac{104}{20} = \dfrac{26}{5} = 5\dfrac{1}{5}$
[2 marks available — 1 mark for converting to improper fractions, 1 mark for completing the calculation]

5 Prime numbers: 2, 3, 5, 7
P(throwing prime number on 6-sided dice) = $\dfrac{3}{6} = \dfrac{1}{2}$
P(throwing prime number on 10-sided dice) = $\dfrac{4}{10} = \dfrac{2}{5}$
$\dfrac{1}{2} \times 300 = 150$, $\dfrac{2}{5} \times 200 = 80$
So an estimated $150 + 80 = 230$ prime numbers will be rolled.
[2 marks available — 1 mark for at least one probability correct, 1 mark for the correct answer]

6 a) $2x^2 - 8x + 5 = 2(x^2 - 4x) + 5$
$= 2((x - 2)^2 - 4) + 5$
$= 2(x - 2)^2 - 8 + 5$
$= 2(x - 2)^2 - 3$, so $a = 2$, $b = -2$ and $c = -3$.
[2 marks available — 1 mark for the correct values of both a and b, 1 mark for the correct value of c]

b) $2x^2 - 8x + 5 = 2(x - 2)^2 - 3$, so solve $2(x - 2)^2 - 3 = 0$.
$(x - 2)^2 = \dfrac{3}{2}$ *[1 mark]*
$x - 2 = \pm\sqrt{\dfrac{3}{2}}$
$x = 2 \pm \sqrt{\dfrac{3}{2}}$ $\left(= 2 \pm \dfrac{\sqrt{6}}{2}\right)$ *[1 mark]*
[2 marks available in total — as above]

7 $\dfrac{3.2 \times 10^4}{8 \times 10^{-k}} = \dfrac{32 \times 10^3}{8 \times 10^{-k}}$
$= \dfrac{32}{8} \times \dfrac{10^3}{10^{-k}} = 4 \times 10^3 \times 10^k = 4 \times 10^{(3 + k)}$
[2 marks available — 2 marks for the correct answer, otherwise 1 mark for any answer in the form 4×10^n]
Alternatively you could have found that $(3.2 \times 10^4) \div (8 \times 10^{-k})$ = $0.4 \times 10^{(4 + k)}$ and then converted this answer into standard form.

8 a) Use the formula for the nth term (n^{th} term = $a + (n-1)d$)
      to make two equations: $a + 2d = 18$ and $a + 6d = 38$
      Subtracting the two equations gives $4d = 20$, so $d = 5$.
      Substitute this value for d back into the first equation:
      $a + 10 = 18$, so $a = 8$
      So the nth term = $8 + (n-1) \times 5 = 5n + 3$
      [3 marks available — 3 marks for the correct answer, otherwise 2 marks for an incorrect expression with the correct coefficient on n or 1 mark for a correct attempt to find the common difference]

   b) Substitute the values of a and d into the formula for the sum of the first 50 terms of a series:
      $\frac{50}{2}[2 \times 8 + (50-1) \times 5] = 25(16 + 245) = 6525$
      [2 marks available — 1 mark for correctly substituting values into a formula for the sum of the first n terms of an arithmetic series, 1 mark for the correct answer]
      Alternatively, you could work out the 50th term using the expression you found in part (a) — it's 253. Then you can use the formula $\frac{n}{2}(a + \text{last term})$.

9 a) (i) $g(-2) = (-2)^2 - 3 = 4 - 3 = 1$ *[1 mark]*
      (ii) $x^2 \geq 0$, so $g(x) \geq -3$ *[1 mark]*

   b) Set $x = 3y - 5$, and rearrange to make y the subject.
      $3y = x + 5$
      $y = \frac{x+5}{3}$
      So $f^{-1}(x) = \frac{x+5}{3}$
      [2 marks available — 1 mark for the correct method, 1 mark for the correct answer]

   c) $gf(x) = g(3x - 5)$
      $g(3x - 5) = (3x - 5)^2 - 3$ *[1 mark]*
      $= 9x^2 - 30x + 25 - 3 = 9x^2 - 30x + 22$ *[1 mark]*
      [2 marks available in total — as above]

10 $k + 2 = \sqrt[3]{\frac{1}{j + l^2}}$
   $(k + 2)^3 = \frac{1}{j + l^2}$ *[1 mark]*
   $j + l^2 = \frac{1}{(k+2)^3}$ *[1 mark]*
   $l^2 = \frac{1}{(k+2)^3} - j$ *[1 mark]*
   $l = \sqrt{\frac{1}{(k+2)^3} - j}$ *[1 mark]*
   [4 marks available in total — as above]

11 The gradient of the line from (2, 7) to (5, 13) = $\frac{13-7}{5-2} = \frac{6}{3} = 2$.
   Two lines are perpendicular if their gradients multiply to equal -1, so a perpendicular line to 2 will have a gradient of $-\frac{1}{2}$.
   The equation $2y = 13 - x$ can be rearranged to give $y = \frac{13}{2} - \frac{1}{2}x$.
   The gradient is $-\frac{1}{2}$ and so the two lines are perpendicular.
   [3 marks available — 1 mark for finding the gradient of the line through (2, 7) and (5, 13), 1 mark for finding the gradient of the perpendicular to this line, 1 mark for showing clearly that the gradient of 2y = 13 – x is equal to this]

12 Label CD as the height x, and AD as length y.
   $x = 6\tan 34° = 4.0470...$ cm *[1 mark]*
   Use Pythagoras' theorem to find y:
   $4.0470...^2 + y^2 = 5.4^2$, so $y^2 = 5.4^2 - 4.0470...^2 = 12.7813...$
   $y = \sqrt{12.7813...} = 3.5751...$ cm *[1 mark]*
   Area of triangle $ABC = \frac{1}{2} \times (3.5751... + 6) \times 4.0470...$ *[1 mark]*
   $= 19.3754...$ cm² $= 19.38$ cm² (2 d.p.) *[1 mark]*
   [4 marks available in total — as above]

13 a) $n(S \cap G) = 6$, so 6 students study Spanish and German.
      Since 4 students study all three languages,
      $6 - 4 = 2$ students study just Spanish and German.
      $n(S) = 27$, so 27 students in total study Spanish.
      So $27 - 8 - 4 - 2 = 13$ students study just Spanish.
      There are 50 students altogether, so
      $50 - (13 + 8 + 4 + 2 + 11 + 3 + 3) = 6$ students study none of the languages.

      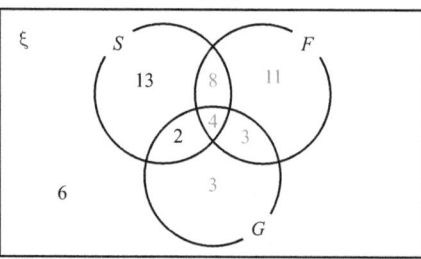

      [2 marks available — 2 marks for all three correct values, otherwise 1 mark for at least one correct value]

   b) $A \subset F$ means A is a subset of F. $A \cap G = \emptyset$ means the intersection of A and G is empty. So A will be at its largest if it includes all students who study French but not German. This is $11 + 8 = 19$ students.
      [2 marks available — 2 marks for the correct answer, otherwise 1 mark for a sum involving 11 or 8]

   c) $13 + 3 + 11 = 27$ students study one of the languages. *[1 mark]*
      So the probability that one is selected at random is $\frac{27}{50}$ (or 0.54) *[1 mark]*
      [2 marks available in total — as above]

   d) $8 + 11 + 4 + 3 = 26$ students study French. *[1 mark]*
      Of these students, 8 study Spanish and 3 study German.
      So the probability is $\frac{8+3}{26} = \frac{11}{26}$ *[1 mark]*
      [2 marks available in total — as above]

14 a) Read across to the graph from a cumulative frequency of $60 \div 2 = 30$. So an estimate of the median length of the songs is 186 seconds.
      [1 mark for an answer in the interval 185 to 186 seconds]

   b) Read across from 15 to find the lower quartile at 172.
      Read across from 45 to find the upper quartile at 204.
      So the interquartile range is $204 - 172 = 32$ seconds.
      [2 marks available — 1 mark for one quartile correct and an attempt to subtract, 1 mark for an answer in the interval 30 to 34 seconds]

15 a)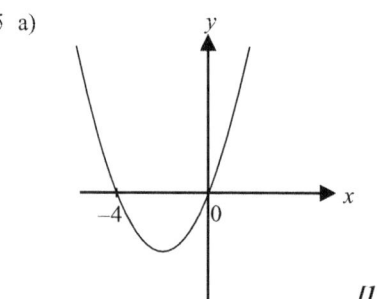

      [1 mark]

   b) The graph is a reflection of $y = f(x)$ in the x-axis, so the equation of the curve is $y = -f(x)$. *[1 mark]*

16 a) The approximated 32 000 is 107% of the original value, so $\frac{32\,000}{107} \times 100 = 29\,906.542...$ hours were spent in meetings in 2018. The difference is $32\,000 - 29\,906.542...$ $= 2093.457...$ hours = 2090 hours (3 s.f.)
[3 marks available — 1 mark for using 107 or 1.07, 1 mark for the correct number of hours for 2018, 1 mark for the correct answer]

b) Find the sum of the upper bounds for 2018 and 2019.
17 000 to the nearest 1000 means the upper bound for the number of hours spent in meetings in 2019 was 17 500.
To find the upper bound for 2018, use the minimum percentage difference. 3% to the nearest percentage point means the lower bound for the percentage is 2.5%.
2.5% fewer hours means the 2018 number is 100% – 2.5% = 97.5% of the 2019 number.
17 500 × 0.975 = 17 062.5
17 500 + 17 062.5 = 34 562.5
So the upper bound for the value of n is 34 562.5 hours.
[4 marks available — 1 mark for using 17 500, 1 mark for using 2.5% or 0.975, 1 mark for finding the correct number for 2018, 1 mark for the correct answer]

17 a) Write the equation in the form $y = mx + c$:
$4y + 2x = 3$
$4y = -2x + 3$
$y = -\frac{1}{2}x + \frac{3}{4}$, so the gradient of the new line is also $-\frac{1}{2}$.
The line will go through (8, 0). Then the line is $y - 0 = -\frac{1}{2}(x - 8)$, which gives $y = -\frac{1}{2}x + 4$.
So the line also goes through (0, 4).
See the line labelled $y = f(x)$ on the grid below.
[2 marks available — 2 marks for the correct line drawn, otherwise 1 mark for a line with the correct gradient]

b) The line in part (a) is $y = -\frac{1}{2}x + 4$, so $f(x) = -\frac{1}{2}x + 4$.
Work out which side of each line should be shaded:
$y \leq f(x)$: $x = 0, y = 0$ gives $0 \leq 4$, which is true, so shade the side containing the origin.
$x \geq 1$: Shade the side where x is greater than 1, which is the side that doesn't contain the origin.
$y \geq x - 1$: $x = 0, y = 0$ gives $0 \geq -1$, which is true, so shade the side containing the origin.

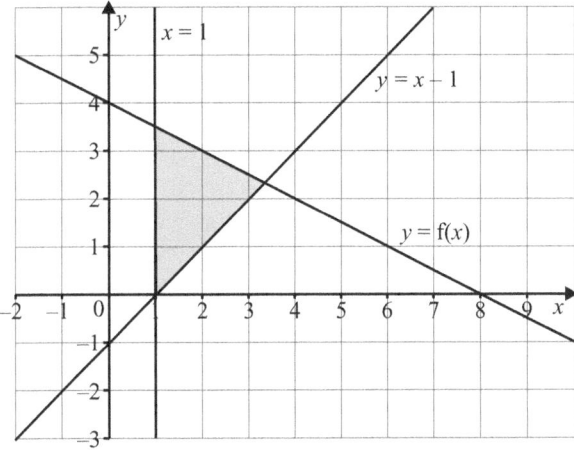

[3 marks available — 3 marks for the correct region shaded, otherwise 1 mark for both x = 1 and y = x – 1 drawn correctly or 2 marks for the lines drawn correctly and a region shaded that satisfies two of the inequalities]

18 The volumes are in the ratio 1:8 so the side lengths are in the ratio $\sqrt[3]{1} : \sqrt[3]{8} = 1:2$ *[1 mark]*
Let R be the radius of the large sphere.
$28 = 4\pi R^2$, so $R^2 = \frac{28}{4\pi}$ *[1 mark]*
$R = \sqrt{\frac{7}{\pi}} = 1.4927...$ cm *[1 mark]*
$r = 1.4927... \div 2 = 0.7463... = 0.75$ cm (2 d.p.) *[1 mark]*
[4 marks available in total — as above]
There are a few ways to answer this question (e.g. you could start by using the ratio 1:4 to find that the surface area of the smaller sphere is 7 cm². You'll still pick up the marks so long as you show your working and get to the right answer.

19 $c + 3 : a + 3 = 2 : 3$
$\frac{c+3}{a+3} = \frac{2}{3}$, so $3(c + 3) = 2(a + 3)$
$3c + 9 = 2a + 6$
$3c - 2a = -3$ *[1 mark]*
$c + 1 : a + 5 = 1 : 2$
$\frac{c+1}{a+5} = \frac{1}{2}$, so $2(c + 1) = a + 5$
$2c + 2 = a + 5$
$2c - a = 3$ *[1 mark]*
$2c - a = 3 \xrightarrow{\times 2} 4c - 2a = 6$ *[1 mark]*

$\begin{array}{r} 3c - 2a = -3 \\ - \quad 4c - 2a = 6 \\ \hline -c = -9 \end{array}$

$2c - a = 3$
$2 \times 9 - a = 3$
$18 - a = 3$
$c = 9$ *[1 mark]* $\quad a = 15$ *[1 mark]*

The ratio $c : a = 9 : 15 = 3 : 5$ in its lowest terms. *[1 mark]*
[6 marks available in total — as above]

20 Angle ABD = angle DAF = 40° *[1 mark]*
(alternate segment theorem) *[1 mark]*
Angle ADB = angle ACB = 63° *[1 mark]*
(angles in the same segment are equal) *[1 mark]*
Angle BAD = 180° – 63° – 40° = 77°
(angles in triangle ABD add up to 180°)
Angle FAO = 90° (angle subtended at the circumference by a diameter is a right angle) *[1 mark]*
Angle BAO = angle BAD – angle DAO
= 77° – (90° – 40°) = 27° *[1 mark]*
[6 marks available in total — as above]

21 Differentiate to find the gradient at x:
$\frac{dy}{dx} = -5x^2 - 2x + 3$
This is positive when $-5x^2 - 2x + 3 > 0$
or $5x^2 + 2x - 3 < 0$.
Find the critical values of this quadratic inequality by solving $5x^2 + 2x - 3 = 0$.
$(5x - 3)(x + 1)$
$5x - 3 = 0$, so $x = \frac{3}{5}$. $x + 1 = 0$, so $x = -1$.

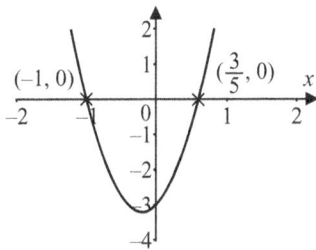

So the graph of $y = -\frac{5}{3}x^3 - x^2 + 3x - 6$ has a positive gradient for $-1 < x < \frac{3}{5}$.

[6 marks available — 1 mark for two terms in the derivative correct, 1 mark for a fully correct derivative, 1 mark for a correct quadratic inequality, 1 mark for a correct method to solve the quadratic, 1 mark for both correct solutions to the quadratic, 1 mark for the correct values of x]

22 $\vec{BC} = 0.25\vec{AD} = 0.25 \times \begin{pmatrix} 16 \\ -12 \end{pmatrix} = \begin{pmatrix} 4 \\ -3 \end{pmatrix}$ *[1 mark]*

$\vec{BE} = \vec{BC} + \vec{CE} = \begin{pmatrix} 4 \\ -3 \end{pmatrix} + \begin{pmatrix} 0 \\ 6 \end{pmatrix} = \begin{pmatrix} 4 \\ 3 \end{pmatrix}$ *[1 mark]*

ABEF is a parallelogram, so $\vec{BE} = \vec{AF}$. Also, $\vec{FB} = -\vec{BF}$.

So $\vec{AB} = \vec{AF} + \vec{FB} = \begin{pmatrix} 4 \\ 3 \end{pmatrix} + \begin{pmatrix} 6 \\ -6 \end{pmatrix} = \begin{pmatrix} 10 \\ -3 \end{pmatrix}$ *[1 mark]*

$|\vec{AB}| = \sqrt{10^2 + (-3)^2}$ *[1 mark]*
$= \sqrt{109}$ *[1 mark]*

[5 marks available in total — as above]
There are many different ways to complete this question. If you used a correct method and labelled your vectors you'll still get the marks.

23 a) Find *AF*, *FC* and *AC* using Pythagoras' theorem.
$AF = \sqrt{6^2 + 8^2} = 10$ cm *[1 mark]*
$FC = \sqrt{6^2 + 12^2} = \sqrt{180}$ cm or $6\sqrt{5}$ cm *[1 mark]*
$AC = \sqrt{8^2 + 12^2} = \sqrt{208}$ cm or $4\sqrt{13}$ cm *[1 mark]*
Use the cosine rule to work out the size of angle *AFC*.

$\cos AFC = \frac{(AF)^2 + (FC)^2 - (AC)^2}{2(AF)(FC)}$

$= \frac{10^2 + (\sqrt{180})^2 - (\sqrt{208})^2}{2(10)(\sqrt{180})}$ *[1 mark]*

$= \frac{3}{5\sqrt{5}}$

$= 0.2683...$

So angle $AFC = \cos^{-1} 0.2683...$
$= 74.435...°$
$= 74.4°$ (1 d.p.) *[1 mark]*

[5 marks available in total — as above]

b) Use the sine rule to work out the size of angle *EHJ*.
$\frac{12}{\sin 44°} = \frac{5}{\sin EHJ}$ *[1 mark]*

$\sin EHJ = \frac{5}{12} \sin 44° = 0.289...$ *[1 mark]*

$EHJ = \sin^{-1} 0.289... = 16.824...° = 16.8°$ (1 d.p.) *[1 mark]*

[3 marks available in total — as above]

Set 2 Paper 2

1 1 kg = 1 000 000 mg = 10^6 mg
2.5 mg = $2.5 \div 10^6$ *[1 mark]* = 2.5×10^{-6} kg *[1 mark]*
[2 marks available in total — as above]

2 a) $20x^3y + 8xy^2 = 4xy(5x^2 + 2y)$ *[1 mark]*

b) $\frac{3x+4}{2} = \frac{5x+3}{3}$
$3(3x+4) = 2(5x+3)$ *[1 mark]*
$9x + 12 = 10x + 6$ *[1 mark]*
$12 = x + 6$
$x = 6$ *[1 mark]*
[3 marks available in total — as above]

c) $2x - 5 < 5x + 4$
$-5 < 3x + 4$
$-9 < 3x$ *[1 mark]*
$-3 < x$ (or $x > -3$) *[1 mark]*
[2 marks available in total — as above]

3 a) See shape **S** on the diagram below.
[2 marks available — 1 mark for a shape drawn with correct scale factor of $\frac{1}{2}$ (in any position), 1 mark for a shape drawn as a correct enlargement from centre (5, 7) (with any scale factor)]

b) See shape **U** on the diagram below.
[2 marks available — 2 marks for the correct final shape, otherwise 1 mark for the correct intermediate translation or for working out $\begin{pmatrix} 6 \\ 1 \end{pmatrix}$ as the single translation vector]

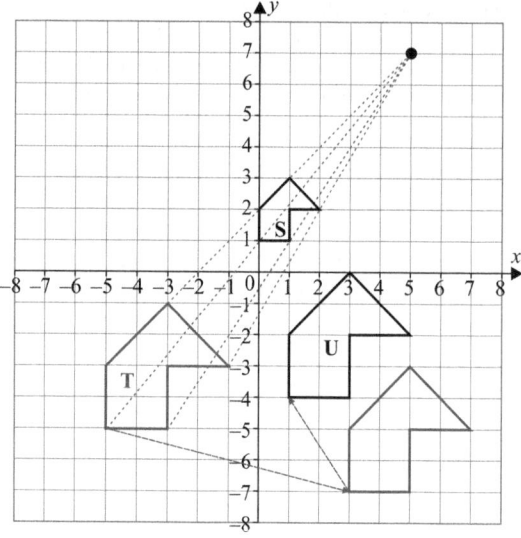

4 a) $392 = 2 \times 2 \times 2 \times 7 \times 7 = 2^3 \times 7^2$
[3 marks available — 1 mark for $2 \times 2 \times 2$, 1 mark for 7×7, 1 mark for $2^3 \times 7^2$]
If you're struggling to find prime factors, try using a factor tree.

b) The common prime factors of *A* and *B* are 2 and 7, so the highest common factor is $2 \times 7 = 14$. *[1 mark]*

c) $A = 2 \times 3 \times 3 \times 7$ so to make this a square number you need to multiply by 2×7 to give:
$2 \times 2 \times 3 \times 3 \times 7 \times 7 = (2 \times 3 \times 7)^2$.
So $k = 2 \times 7 = 14$ *[1 mark]*

5 $12 \leq m < 15$ bar has height 8 and frequency $3 \times 8 = 24$.
   $20 \leq m < 25$ bar has height 3.6 and frequency $5 \times 3.6 = 18$.

Time (m) in minutes	Frequency
$10 \leq m < 12$	8
$12 \leq m < 15$	24
$15 \leq m < 20$	30
$20 \leq m < 25$	18
$25 \leq m < 40$	30

$10 \leq m < 12$ bar has width $12 - 10 = 2$ and height $= \frac{8}{2} = 4$.

$25 \leq m < 40$ bar has width $40 - 25 = 15$ and height $= \frac{30}{15} = 2$.

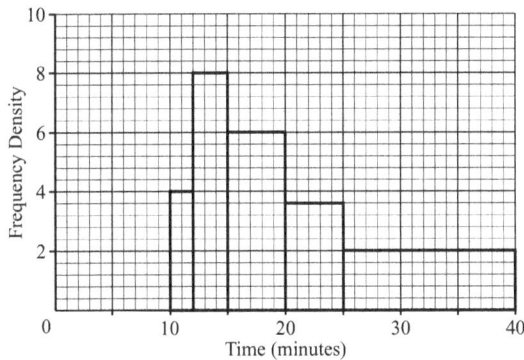

[4 marks available — 1 mark for each correct entry in the frequency table, 1 mark for each bar drawn correctly on the histogram]

6 a) x = exterior angle $= \frac{360°}{16} = 22.5°$ *[1 mark]*

   b)
   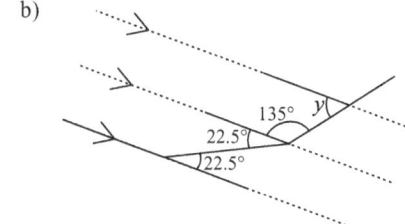

   Angle in z-shape = 22.5° (alternate angles)
   Interior angle = 180° − 22.5° = 157.5°
   157.5° − 22.5° = 135°
   y = 180° − 135° = 45° (allied angles)
   [2 marks available — 2 marks for correct answer, otherwise 1 mark for attempt at using angle rules or finding the interior angle]

7 Multiply two brackets out first and then multiply the product by the third bracket.
   $(2u + 5)(u + 1) = 2u^2 + 2u + 5u + 5 = 2u^2 + 7u + 5$
   $(2u^2 + 7u + 5)(u − 10) = 2u^3 − 20u^2 + 7u^2 − 70u + 5u − 50$
   $= 2u^3 − 13u^2 − 65u − 50$
   [3 marks available — 1 mark for multiplying two pairs of brackets correctly, 1 mark for multiplying three pairs of brackets correctly, 1 mark for simplifying correctly]

8 $m\mathbf{a} + n\mathbf{b} = m\begin{pmatrix} 2 \\ -1 \end{pmatrix} + n\begin{pmatrix} 5 \\ 3 \end{pmatrix} = \begin{pmatrix} 2m + 5n \\ -m + 3n \end{pmatrix} = \begin{pmatrix} 1 \\ -6 \end{pmatrix}$

   so $2m + 5n = 1$ and $-m + 3n = -6$ *[1 mark]*

   $-m + 3n = -6 \xrightarrow{\times 2} -2m + 6n = -12$ *[1 mark]*

   $-2m + 6n = -12$
   $\underline{2m + 5n = 1} \quad +$
   $11n = -11$
   $n = -1$ *[1 mark]*

   $-m + (3 \times -1) = -6$
   $-m - 3 = -6$
   $-m = -3$
   $m = 3$ *[1 mark]*

   [4 marks available in total — as above]

9 a) P(scores 1st) = 0.7
   P(scores both) = P(scores 1st) × P(scores 2nd given scores 1st)
   So P(scores 2nd given scores 1st)
     = P(scores both) ÷ P(scores 1st) = 0.56 ÷ 0.7 = 0.8
   P(misses 2nd given scores 1st) = 1 − 0.8 = 0.2
   P(misses 1st) = 1 − 0.7 = 0.3
   P(misses both) = P(misses 1st) × P(misses 2nd given misses 1st)
   So P(misses 2nd given misses 1st)
     = P(misses both) ÷ P(misses 1st) = 0.18 ÷ 0.3 = 0.6
   P(scores 2nd given misses 1st) = 1 − 0.6 = 0.4

   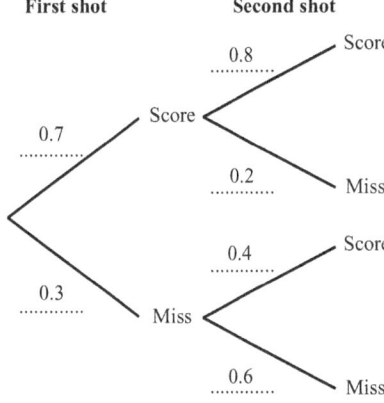

   [3 marks available — 1 mark for both probabilities of his first shot, 1 mark for both probabilities for second shot given he scored the first, 1 mark for both probabilities for second shot given he missed the first]

   b) P(misses 1st but scores 2nd) = 0.3 × 0.4 = 0.12 *[1 mark]*
   P(scores 1st but misses 2nd) = 0.7 × 0.2 = 0.14 *[1 mark]*
   Jack is more likely to miss with his second shot. *[1 mark]*
   [3 marks available in total — as above]

10 $\frac{\sqrt{32} + \sqrt{18}}{\sqrt{2} - 1} = \frac{4\sqrt{2} + 3\sqrt{2}}{\sqrt{2} - 1}$

    $= \frac{7\sqrt{2}}{\sqrt{2} - 1}$

    $= \frac{7\sqrt{2}}{\sqrt{2} - 1} \times \frac{\sqrt{2} + 1}{\sqrt{2} + 1}$

    $= \frac{7 \times 2 + 7\sqrt{2}}{2 - \sqrt{2} + \sqrt{2} - 1}$

    $= 14 + 7\sqrt{2}$

    [4 marks available — 1 mark for correctly simplifying two surds, 1 mark for a correct method of rationalising the denominator, 1 mark for each of a and b correct]

11 Opposite angles in a cyclic quadrilateral add up to 180° so
    $5x + 12° + x + 18° = 180°$ *[1 mark]*, which gives $6x + 30° = 180°$,
    and so $6x = 150°$ and $x = 25°$. *[1 mark]*
    Angle $BCD = 5 \times 25° + 12° = 125° + 12° = 137°$.
    $137° + 43° = 180°$ so BCD and ADC are allied angles. *[1 mark]*
    BC is parallel to AD and $ABCD$ is a trapezium. *[1 mark]*
    [4 marks available in total — as above]

12 $c \propto a^2$, so $c = ka^2$ *[1 mark]*
    When $a = 10$, $c = 450$ so $450 = k \times 10^2 = 100k$
    So, $k = 4.5$ *[1 mark]* and $c = 4.5a^2$
    When $c = 882$, $882 = 4.5a^2$ *[1 mark]*
    $a^2 = 196$
    $a = 14$ *[1 mark]*
    [4 marks available in total — as above]

13 a)

x	−1	−0.5	0	0.5	1	1.5	2	2.5
y	0	−1.375	−1	0.375	2	3.125	3	0.875

[2 marks available — 2 marks for all four values correct, otherwise 1 mark for at least two values correct]

b) See the curve labelled $y = -x^3 + 2x^2 + 2x - 1$ on the grid below.
[2 marks available — 1 mark for at least five points from the table plotted correctly, 1 mark for a correct curve]

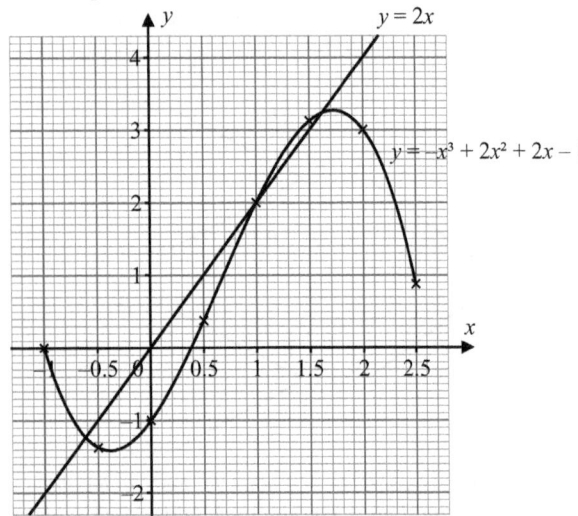

c) Add terms to both sides of the equation so that the left-hand side becomes the equation of the curve you've just drawn:
$-x^3 + 2x^2 = 1$
$-x^3 + 2x^2 + 2x - 1 = 1 + 2x - 1$
$= 2x$
So draw $y = 2x$ (see the grid above) and find the values of x at the points where it intersects the curve.
$x = -0.6$, $x = 1$ and $x = 1.6$ (1 d.p.)
[4 marks available — 1 mark for equating $-x^3 + 2x^2 + 2x - 1$ and $2x$, 1 mark for drawing $y = 2x$ on the grid, 1 mark for a correct x-value in the interval −0.7 to −0.5, 0.9 to 1.1 or 1.5 to 1.7, 1 mark for the remaining two correct x-values in the allowed ranges]

14 Call the shortest sides s.
$\tan 29° = \dfrac{s}{16.7}$, so $s = 16.7 \tan 29° = 9.25...$ cm.
Call the remaining side t, then use Pythagoras' theorem.

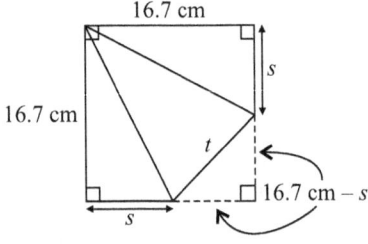

$t^2 = (16.7 - 9.25...)^2 + (16.7 - 9.25...)^2 = 110.79...$
$t = \sqrt{110.79...} = 10.52...$ cm
Perimeter $= 16.7 + s + t + s + 16.7$
$= (2 × 16.7) + (2 × 9.25...) + 10.52... = 62.43...$ cm
$= 62.4$ cm (1 d.p.)
[5 marks available — 1 mark for correctly substituting into a trig ratio, 1 mark for correctly finding any missing side in the right-angled triangle, 1 mark for a correct method to determine the length of the remaining missing side in the shape, 1 mark for the correct length of the remaining missing side, 1 mark for the correct answer]
Alternatively, you could find the hypotenuse of the right-angled triangle and work out the size of the angle opposite t. Then you can use the cosine rule to find t.

15 Area of circle $= \pi r^2$
Area of entire circle $= \pi × 11^2 = 121\pi$
Area of inner two circles (grey and white) $= \pi × (5 + 2)^2 = 49\pi$
Area of innermost grey circle $= \pi × 5^2 = 25\pi$
Shaded area of entire circle $= 121\pi - 49\pi + 25\pi = 97\pi$
Area of shaded sector $= \dfrac{120}{360} × 97\pi = \dfrac{1}{3} × 97\pi$
$= 101.578$ cm² (3 d.p.)
[4 marks available — 1 mark for finding the area of the entire circle, 1 mark for finding the area of one of the inner circles, 1 mark for the shaded area of the entire circle, 1 mark for the correct answer]
You could also solve this problem without finding the areas of the entire circles by just working with the areas of the sectors instead. You'll still get the marks if you do this and get the correct answer.

16 $6x - 3$ factorises to $3(2x - 1)$ and
$2x^2 + 7x - 4$ factorises to $(x + 4)(2x - 1)$. *[1 mark]*
$x^2 - 16 = (x - 4)(x + 4)$ *[1 mark]*
So, $\dfrac{6x - 3}{2x^2 + 7x - 4} \div \dfrac{15}{x^2 - 16}$
$= \dfrac{3(2x - 1)}{(x + 4)(2x - 1)} \div \dfrac{15}{(x - 4)(x + 4)}$
$= \dfrac{3(2x - 1)}{(x + 4)(2x - 1)} × \dfrac{(x - 4)(x + 4)}{15}$ *[1 mark]*
$= \dfrac{3(2x-1)}{(x+4)(2x-1)} × \dfrac{(x-4)(x+4)}{15}$
$= \dfrac{3(x - 4)}{15} = \dfrac{x - 4}{5}$ *[1 mark]*
[4 marks available in total — as above]

17 If r is the radius and h is the height of the small cone, then the full cone will have height $2h$ and radius $2r$. *[1 mark]*
Volume of small cone $= \dfrac{1}{3}\pi r^2 h$
Volume of full cone $= \dfrac{1}{3}\pi(2r)^2 × 2h = \dfrac{8}{3}\pi r^2 h$ *[1 mark]*
Volume of frustum $= \dfrac{8}{3}\pi r^2 h - \dfrac{1}{3}\pi r^2 h = \dfrac{7}{3}\pi r^2 h$ *[1 mark]*
Ratio of small cone : frustum $= \dfrac{1}{3}\pi r^2 h : \dfrac{7}{3}\pi r^2 h = 1 : 7$ *[1 mark]*
[4 marks available in total — as above]

18 a) A decrease of 24% means the height is 100% − 24% = 76%
$= 0.76$ *[1 mark]* of the height before the bounce.
Using the compound decay formula:
$140 × 0.76^5$ *[1 mark]* $= 35.4973... = 35.5$ cm (1 d.p.) *[1 mark]*
[3 marks available in total — as above]
Alternatively, you could have multiplied 140 by 0.76, then the answer of this by 0.76, and again and again and again.

b) 178.5 cm is 170% of the height on the second try.
Height on second try $= \dfrac{178.5}{170} × 100 = 105$ cm
The difference between the height of the first try and the second try is $140 - 105 = 35$ cm.
So $P = \dfrac{\text{difference}}{\text{original}} × 100 = \dfrac{35}{140} × 100 = 25$
[4 marks available — 1 mark for using 170 or 1.7, 1 mark for finding 105 cm, 1 mark for a correct method to find P using the height of the second try, 1 mark for the correct answer]
Instead of subtracting to find 35, you could divide 105 cm by 140 cm and find that the height of the second try is 75% of the height of the first try. So P% = 100% − 75% = 25%. You'll still get the marks if you use this or any other correct method.

19 a) For the product to be odd, both cards must be odd.
All prime numbers other than 2 are odd,
so 11 − 4 = 7 of the cards are odd.
P(odd first card) = $\frac{7}{11}$ *[1 mark]*
P(odd second card) = $\frac{6}{10} = \frac{3}{5}$ *[1 mark]*
Use the AND rule: P(odd product) = $\frac{7}{11} \times \frac{3}{5} = \frac{21}{55}$ *[1 mark]*
[3 marks available in total — as above]

b) For the product to be 0, at least one of the cards must be 0.
P(at least one 0) = 1 − P(no 0s) *[1 mark]*
Use the AND rule:
P(no 0s) = P(first card not 0) × P(second card not 0)
= $\frac{11}{12} \times \frac{11}{12} = \frac{121}{144}$ *[1 mark]*
So P(at least one 0) = 1 − $\frac{121}{144} = \frac{23}{144}$ *[1 mark]*
[3 marks available in total — as above]
Alternatively, you could consider all of the possible ways of choosing at least one 0 and add up their probabilities.

20 Gradient of OP = $\frac{-6-0}{-8-0} = \frac{-6}{-8} = \frac{3}{4}$ *[1 mark]*
Gradient of tangent is the negative reciprocal, which is $-\frac{4}{3}$. *[1 mark]*
$y = mx + c$ where m is the gradient, and c is the y-intercept:
so $y = -\frac{4}{3}x + c$ *[1 mark]*
Substitute in point P to find c:
$-6 = -\frac{4}{3}(-8) + c$
$-6 = \frac{32}{3} + c$, so $c = -\frac{50}{3}$
The equation of line l is $y = -\frac{4}{3}x - \frac{50}{3}$ *[1 mark]*
[4 marks available in total — as above]

21 Acceleration = $\frac{dv}{dt} = 3t^2 + 4wt + w^2$ *[1 mark]*
When $t = 2$, $\frac{dv}{dt} = 10$ m/s²
$3(2)^2 + 4(2)w + w^2 = 10$
$12 + 8w + w^2 = 10$
$w^2 + 8w + 2 = 0$ *[1 mark]*
$w = \frac{-8 \pm \sqrt{8^2 - 4 \times 1 \times 2}}{2 \times 1}$ *[1 mark]*
$= \frac{-8 \pm \sqrt{56}}{2}$
$= \frac{-8 \pm 2\sqrt{14}}{2}$
$= -4 \pm \sqrt{14}$ *[1 mark]*
[4 marks available in total — as above]
Instead of the using the quadratic formula, you could have completed the square: $w^2 + 8w + 2 = (w + 4)^2 - 14 = 0$. So $(w + 4)^2 = 14$, then $w + 4 = \pm\sqrt{14}$ and $w = -4 \pm \sqrt{14}$

22 The perimeter = 1, so $2\left(\frac{1}{3x} + \frac{x}{6}\right) = 1$
So, $2\left(\frac{1}{3x} + \frac{x}{6}\right)$ *[1 mark]*
$= 2\left(\frac{3x^2 + 6}{18x}\right) = 6\left(\frac{x^2 + 2}{18x}\right) = \frac{x^2 + 2}{3x} = 1$ *[1 mark]*
$x^2 + 2 = 3x$
$x^2 - 3x + 2 = 0$
$(x - 1)(x - 2) = 0$ *[1 mark]*, so $x = 1$ or $x = 2$ *[1 mark]*
Substituting $x = 2$: $AB = \frac{1}{6}$ and $BC = \frac{1}{3}$,
but AB is supposed to be longer than BC, so $x \neq 2$.
Substituting $x = 1$: $AB = \frac{1}{3}$ miles and $BC = \frac{1}{6}$ miles *[1 mark]*
[5 marks available in total — as above]

CGP

www.cgpbooks.co.uk